WHAT PRICE FOOD?

Also by Paul Streeten

DEVELOPMENT PERSPECTIVES
FIRST THINGS FIRST (*with Shahid Javed Burki, Mahub ul Haq, Norman Hicks and Frances Stewart*)
FOREIGN INVESTMENT, TRANSNATIONALS AND DEVELOPING COUNTRIES (*with Sanjaya Lall*)
THE FRONTIERS OF DEVELOPMENT STUDIES
ECONOMIC INTEGRATION
RECENT ISSUES IN WORLD DEVELOPMENT (*editor with Richard Jolly*)
TRADE STRATEGIES FOR DEVELOPMENT (*editor*)
UNFASHIONABLE ECONOMICS (*editor*)
AID TO AFRICA
DIVERSIFICATION AND DEVELOPMENT: The Case of Coffee (*with Diane Elson*)
VALUE IN SOCIAL THEORY (*editor*)
THE CRISIS OF INDIAN PLANNING (*editor with Michael Lipton*)
COMMONWEALTH POLICY IN A GLOBAL CONTEXT (*editor with Hugh Corbet*)

Edited by *Sanjaya Lall and Frances Stewart*

THEORY AND REALITY IN DEVELOPMENT: Essays in Honour of Paul Streeten

What Price Food?

Agricultural Price Policies in Developing Countries

Paul Streeten

Foreword by Michael Lipton

Cornell University Press   Ithaca, New York

Cornell Paperbacks edition first published 1988 by Cornell University Press, by arrangement with St. Martin's Press, Inc.

Library of Congress Cataloging-in-Publication Data

Streeten, Paul.
  What price food?

  Bibliography: p.
  Includes index.
  1. Agricultural prices—Government policy—
Developing countries. 2. Food prices—Government
policy—Developing countries. I. Title.
[HD1417.S73 1988]   388.1′8   88-47773
ISBN 0-8014-9533-4 (pbk. : alk. paper)

Printed in the United States of America

The paper in this book is acid-free and meets the guidelines for permanence and durability of the Committee on Production Guidelines for Book Longevity of the Council on Library Resources.

Contents

Acknowledgements

I am grateful to Sara Berry, David Davies, Susan George, Nurul Islam, Leslie Helmers, A. R. Khan, Robert Klitgaard, Mohan Rao, Shlomo Reutlinger, Hans Singer, and Peter Timmer for helpful comments, and to Michael Lipton for encouragement. I am also indebted to Leslie Helmers and the Studies Unit of the Economic Development Institute of the World Bank, and Christopher Willoughby, Director of the Economic Development Institute, for having given me the opportunity and the support to write this book.

PAUL STREETEN

Foreword

A spectre is haunting Africa – the spectre of pricism. This clear-headed, undogmatic book exorcises that spectre. In so doing, it rescues price policy from the excessive enthusiasm of the anti-state ideologists. For Streeten shows that price adjustments can succeed only if backed, not by less state action as the pricists claim, but by more.

Pricism is the belief that 'getting prices right' – largely by minimising state involvement in the pricing and marketing of outputs, inputs, and foreign exchange – is a comprehensible, attainable and optimal way to solve the problems of poor countries, especially in food and agriculture. The deep political compulsions underlying urban bias, compulsions brilliantly analysed on pp. 77–84 below, are conveniently reduced by pricists to economic mistakes in managing food prices or foreign exchange-rates.

Pricism stalks Africa in particular, because that continent's policies have been most harmed by extreme urban bias. Hence genuine bad luck, with export prices and drought and debt, in the early 1980s found many countries so weakened as to have no options; pricist groups in aid donor agencies were therefore free to turn their views into conditions for (indispensable) cash. Arguably, for exceptional crops in exceptional countries – sisal in Tanzania, cocoa in Ghana, coffee in Uganda – price policy errors had been so extreme, and so hard for the farmer to escape, as to render their correction 'top priority' in policy reform; however, the pricists were wrong to regard those conditions as typical, or as corrigible without a major increase and improvement in attention and outlay by African governments on what Streeten calls the 'other five Ins' of agriculture, apart from price incentives. These are input delivery systems (especially water and irrigation); innovation via crop research; information, via extension; infrastructure (more health care, not airports); and institutional reform, especially of tenure and credit.

Of course governments are not perfect at improving these things. They have their own biases, and face pressures, often malign, from within and without. But they are the only governments available; and private persons or groups can seldom be expected to improve such public goods at all. As Streeten points out, pricists stress that 'Market failure [need] not justify government intervention. But equally, government failure need not justify reliance on the market.'

Streeten's analysis is built around a trilemma. High food prices for farm sellers stimulate food production. Low food prices for poor buyers help them to eat adequately. Both together mean that the government budget, and/or the balance of payments, is overstretched. The trilemma is sharpened, as this book shows, by the many goals of farm policies (contrary to the pricists, price management is only ore policy, and efficiency is only one goal, alongside growth, equity and stability); by a complex history of interventions, such that many apparent 'errors' in price policy exist partly to make up, however imperfectly, for non-price policy biases; and by vested interests behind each established policy decision. The problems can be resolved, but it will require subtle analysis, case-specific, yet respecting the general evidence that as a rule farmers cannot respond to price incentives by raising *total* output substantially, unless appropriate inputs and improved farming methods are available near the margin of profitability. The East Asian evidence, widely misinterpreted as justifying pricism, in fact proves that major public actions in support of rural activity can compensate farmers for extractive price policies (1950–70) – and best support reformed ones (1970–86).

Paul Streeten's clear, learned and witty book clarifies all these issues. It is essential reading for three groups of people. Third World policy-makers will learn to respond to pricist foreigners neither with flat hostility nor with despairing acceptance, but with counter-proposals assuming (to rephrase James Tobin's remark on monetary policy) neither that only prices matter, nor that prices don't matter, but that prices matter *too*. Pricists themselves – at least the majority, who are not pure ideologists of state-hate – will learn from Streeten's thorough, analytical eclecticism to consider politics as well as economics, human ideals as well as human self-interest, and to understand with Matthew Arnold that the doctrine of the One Thing Necessary, *porro unum est necessarium*, is deeply illiberal (and absurd) whether the One Thing is 'get the prices right' or – as per the religious dogmatists mocked by Arnold in the 1860s – 'permit a man to marry his deceased wife's sister'. Finally, all students and scholars of economics and development will re-learn from Streeten that 'rigour' means, not (as Leamer warns) econometrics with the con concealed, but respect for evidence and logic, even when they stem from other people's disciplines or ideological stances.

MICHAEL LIPTON

1 The Dilemma

> Oh! God! that bread should be so dear,
> And flesh and blood so cheap!
> *The Song of the Shirt*, Thomas Hood

Policy-makers in the developing countries are confronted with a fundamental dilemma. On the one hand, they want high prices for food in order to encourage agricultural production. On the other hand, they want low prices to protect (at least in the short run) the poor buyers of food. There is a large literature advocating each of these courses. A heavy weight of writing has now accumulated which shows that agricultural producers have been discriminated against in many low-income countries, and that this has been detrimental to growth, equity, employment, and poverty alleviation. But there is also a large literature showing that adequate supply of food is not enough to eliminate malnutrition, hunger and starvation; poor people must also have 'access' or 'exchange entitlements' to the food. Above all, this means adequate incomes and low prices of food. In addition, they must have access to complementary goods and services, such as education and health services, for which income is not a sufficient condition, and the distribution of food within the household must meet the nutritional requirements of all its members.

It is often said that a secondary dilemma is that between the role of the agricultural sector as a producer of food, which calls for its modernization, and its role as a provider of resources for the rest of the economy, which calls for its exploitation. The resources may be savings for industrialization or foreign exchange or tax revenue.[1] This problem is often wrongly stated, requires a careful analysis, and is briefly discussed later. The main purpose of this book is to sort out the issues underlying these actual and apparent dilemmas and to contribute to the formulation of guidelines for policy-makers. The intended readership of this book are policy-makers in developing countries, and particularly those who wish to embark on reforms and to manage the transition from badly managed to better managed economies sensibly. I hope the book will also prove useful to undergraduate and graduate students of development policy and the wider development community concerned with reducing poverty, hunger and malnutrition.

1

HOW THE DILEMMA AROSE

The dilemma has three origins, each presenting an obstacle to reform. First, agricultural price policies have attempted to pursue many objectives, which will be set out more fully below. It is not only growth with equity, but numerous proximate and intermediate objectives (or constraints) that have dictated policies. Some of these conflict with one another. Where they are potentially compatible, it would be surprising if a single instrument, price policy, were sufficient to achieve all these objectives. The task is to reconcile these objectives, to balance them where they conflict, weighing the trade-offs, and to choose a battery of feasible and acceptable instruments, where they are consistent with one another. By a combination of price policies, taxes and subsidies, exchange rates, tariffs and other measures, a wide range of objectives can, in principle, be achieved. But where there is conflict, there has to be choice. It is the test of a good policy-maker that he can reconcile objectives where they can be made compatible, that he chooses where they conflict, and that he knows which is which.

The second origin of the dilemma is historical. It is easy to set out formally a range of targets and instruments and to proceed to optimize an objective function. Reality is quite different. Policy-makers find themselves with a heritage of a complex structure of interventions relating not only to explicit and implicit measures for output prices, but also with input subsidies attempting to offset the deterrent effects of low output prices, and the high costs of some inputs due to protection, explicit and implicit taxes on the consumption goods bought by farmers, various direct controls and *ad hoc* measures to meet specific pressures, the direction and size of the ultimate impact of which is quite uncertain. A full investigation has to look not only at the prices producers get, but also at the prices they have to pay for inputs and for all other goods they purchase. For in order to get at 'real' producer prices we have to correct nominal money prices for the prices of all the goods farmers spend their receipts on, i.e. the costs of all their inputs and the prices of the consumer goods they buy. General price indexes, which are normally used to deflate producer prices in order to derive the agricultural–industrial terms of trade, often do not reflect correctly the movement of the prices of the specific goods bought by farmers. The next step is to discover the real incomes earned by farmers. To do this we have to allow for changes in agricultural productivity. Even with falling 'real' prices, real incomes may rise if productivity grows. It is possible for the commodity terms of trade to turn against agriculture, while the income terms of trade turn in its favour. In Kenya, for example, producer prices for barley and tobacco fell in real terms in the seventies, but production increased to self-sufficiency by 1983.[2] While the price of food falls, the incomes of food producers can rise. It is in these movements that the ultimate solution of the basic dilemma between higher incomes for producers and lower prices for consumers is to be found. But the path to this ultimate solution, is, as we shall see, strewn with obstacles.

It is true that some of these government interventions sprang from a desire to encourage industrialization by keeping real wages low,[3] and from pessimism about the opportunities of international trade. But they were not necessarily intended to discriminate against agriculture (though they often did), whose contribution to development was acknowledged by subsidies to inputs, technical advisory services, and encouragement of research into new technologies. Some policies attempted to compensate for the harmful impact of policies in other parts of the economy, such as credit subsidies as compensation for overvalued exchange rates. The impact of this mixed heritage of interventions is detrimental to agriculture in most developing countries (the exceptions are some east Asian countries that did not suffer from them), but the size of the deterrent total impact is difficult to assess. It can be argued that there is a political rationale behind the panoply of interventions and even their inefficiencies: to discriminate against the small farmer in favour of the urban population and the large and rich farmer. These issues are briefly discussed in the section on the politics of food pricing. On the other hand, it may be that many of the effects are unintended and the prevailing ignorance about them calls for a careful analysis before judgement can be pronounced.

Even if ignorance and uncertainty were removed, so that the desirable course of action becomes crystal clear, there is a third obstacle to its pursuit. The historical heritage of these measures has left vested interests, political pressures and political constraints, which to the policy-maker are at least as real as the technical constraints are to the engineer or the economic constraints to the economist. Political constraints should be neither ignored nor accepted as ultimate facts. They, too, call for analysis. In this case the problem is how to build a political base for the desirable reforms. To talk of political will is not enough. What is needed is an analysis of the reformist coalitions that would provide the constituency for reform.

In order to resolve the dilemma three things are necessary: first, a clear awareness of the set of objectives and their conflict or consistency, second, an economic analysis and an empirical estimate of the impact of existing measures and of any proposed reforms; and, third, a political analysis of how interests and support can be mobilized for these reforms. It is hoped that this book contributes to all three.

2 Multiplicity of Objectives

The ultimate objective of policy-makers can be loosely described as growth with equity, or the provision of the opportunities for full human and social development. But there are at least five proximate objectives and two or three constraints of food price policies which are listed below.

1. Allocational efficiency, i.e. the objective is to raise food production and productivity at the lowest possible costs and to avoid food shortages.
2. Acceleration of aggregate economic growth through the balanced expansion of agriculture (including non-food crops), manufacturing industry and services. This objective raises the question of the relations between food and the rest of agriculture, and between agriculture, industry, and other sectors, and their respective contribution to economic growth.
3. A range of social objectives (some of which may themselves be in conflict with one another) including the elimination of hunger, starvation, malnutrition and undernutrition, poverty alleviation and income redistribution; special attention to the fate of small farmers, employment creation for landless labourers or the inhabitants of poor regions; reduction in rural–urban income differentials and migration.
4. National food security in the face of international uncertainties. Price stabilization for both consumers and producers, so that their real incomes do not fall below a critical minimum. There are also good macroeconomic reasons for stabilizing agricultural prices.
5. Political objectives such as the avoidance of political disturbances and riots,[1] or the loss of political support from powerful urban groups. This objective comprises both the desire for social and political stability (which are important for investment and economic growth) and the desire of a particular government to stay in power.

The constraints are the budget and the balance of payments, and we might add the constraint set by administrative capacity. The budgetary constraint consists in minimizing the burden on the budget, and on scarce administrative skills. The balance of payments constraint

5

consists in keeping within available foreign exchange resources.[2] The importance of these constraints cannot be overemphasized. In view of the multiple objectives, some of which are in conflict with each other, either something has to be sacrificed, or an additional instrument, such as a tax, or a subsidy, or a control, has to be introduced, in addition to the price policy. And it is the obstacles to their introduction that prevent the achievement of the objectives. It should only be added that the distinction between constraint and objective is to some extent artificial, because it is political priorities that determine the expenditure of actually collected taxes.

In view of these five interrelated objectives (and the two or three constraints), we shall have to consider the use of several instruments besides agricultural price policies. But in order to avoid this book becoming a general essay on development, attention to instruments other than 'getting agricultural prices right' and the objectives they serve will be given only to the extent to which these are relevant to a discussion of agricultural price policies and their aims.

On certain simplifying assumptions there is a greater degree of consistency between these objectives than when certain practical constraints are taken into account. For example, it is possible to combine the objective of higher prices to small farmers, equity in income distribution, and lower prices to poor food consumers by a system of incentive prices, combined with a progressive land tax, an income tax and targeted food subsidies to vulnerable groups of consumers. Or it is possible to combine foreign food aid with encouragement to domestic agriculture by using local counterpart funds from the sale of the food aid at market clearing prices for deficiency payments to farmers who would otherwise be harmed by the food aid. It is also possible to encourage small farmers to adopt new technologies in the face of the greater uncertainty of new technologies by higher supply prices and a crop insurance scheme. When budgetary, political or administrative constraints are taken into account, the question becomes either one of how to overcome the strains of the transition to a more rational system of incentives or, if it is altogether precluded, one of balance and compromise between alternative objectives. It should also be remembered that some of the objectives for which food pricing policy has been enlisted are more effectively pursued by other means, for example a more equal income distribution by land reform and employment creation rather than low food prices.

The budgetary constraint, combined with political pressures, adds a third horn to the dilemma described in the first sentence of this book.

The inability to raise extra revenue, or when revenue is raised, to spend it on the types of subsidy indicated, sharpens the basic dilemma between high and low food prices. 'The problem is you can't afford to keep low prices for the urban consumer and high prices for the farmer,' says Kenneth Shwedel, a senior economist at Banamex, a Mexican Bank. 'If you raise farm prices, you get budgetary problems. But if you raise tortilla prices, you get urban problems.'[3]

The plan of this book is to discuss, after some introductory remarks, these objectives and constraints in turn, with occasional excursions into related topics. After the section on the politics of food pricing, the question of the international implications of reforming policies is taken up. Next, some of the wider issues of eradicating hunger are discussed and the book closes on an agenda of future research.

3 Towards a Country and Crop Typology

The impact of price policies and other measures on the five objectives will vary according to the country and the crop. The distribution of land and tenurial arrangements clearly make a difference to the impact of incentives. Equally clearly, whether crops can be grown additionally to existing ones, and how long they take to mature will make a difference to the impact of price incentives on growing these crops and achieving the objectives. It is possible to take two extreme positions, or an intermediate one. On the one hand, it could be argued that there are certain universal economic principles that apply at all times to all countries and all crops. On the other hand, it could be argued that each situation is unique and that nothing can be said in general. The position of this book is intermediate between these two extremes. It is possible to identify certain categories of crops and countries to which certain general principles apply, but in view of the complexity of the situations, great caution is necessary and only very rough and ready generalizations are possible. As far as crops are concerned, the following four categories are suggested.

1. Subsistence crops, mainly consumed by those who produce them, or sold by them in local markets: sorghum, millet, cassava, pulses, yams, teff, root crops. To these might be added less preferred grains like maize.
2. Grains sold and bought in national and international markets (as well as consumed on the farm): wheat, rice, maize, barley.
3. Non-grain food cash crops: sugar, vegetable oils, nuts, vegetables.
4. Export non-food cash crops: coffee, tea, cocoa, tobacco, pyrethrum and the agricultural fibres: cotton, sisal, jute, kenaf, rubber.

These non-food cash crops can be an important source of income for small farmers and labourers. But so can many other activities.[1]

In addition to crops, there are livestock products, such as milk, beef, pork, poultry and eggs. These are an important source of food for many rural and urban consumers, but will not be discussed in this book.

In a country typology relevant to a classification for the purpose of assessing the impact of food price policies on the range of objectives, three features will make a difference: available resources, demographic and socio-political characteristics, and the size of the country. The following features and questions will be important.

1. The system of land distribution and the proportion of farmers producing largely for own use in the total agricultural population. South Asia and Africa have many more of these than Latin America.
2. Whether there is pressure on land or whether arable land is in surplus. Bangladesh would be an example of the former, Botswana or Sudan of the latter.
3. Whether most rural households own enough land to feed themselves or whether there are many landless labourers. Africa would illustrate the former, south Asia the latter.
4. Whether the country is a grain exporter or importer or on balance self-sufficient.
5. Whether the country is a major exporter of a competing non-cereal agricultural product, such as jute or sisal or cotton, like Bangladesh, Tanzania and Egypt.
6. Whether the country is rich in natural resources or not.
7. What institutional arrangements for marketing the crops exist: are they public or private; if private, are they monopolistic or competitive?
8. The entrepreneurial ability and responsiveness of farmers.
9. The consumption pattern and eating habits of the people (e.g. wheat or rice or cassava).
10. The skills they have (e.g. ability to bake bread when wheat replaces rice).
11. Whether there are large rural–urban income differentials, which will determine the proportion allocated to food production.
12. The administrative capacity of the country, which will determine the extent to which targeted subsidy or ration programmes can be implemented.

While this list could be extended indefinitely, the most important features of a typology from the point of view of the impact of price policy will refer to the distribution of land ownership (equal or unequal), the method of cultivation (multinational corporations, plantations, large commercial farms, cooperatives, small family farms), the existence of landless labourers and their access to education, credit, water and markets, the efficiency of the marketing institutions, and the administrative capacity of the government.

Finally, the implementation of non-price policies by the government will have an important influence on the effectiveness of price policies. These non-price policies relate, depending upon the other features discussed above, to agricultural technology (especially new varieties and irrigation), to transport and other physical infrastructure, to the supply of agricultural inputs such as fertilizer, and consumer goods on which the farmers' incomes are spent, to institutions such as banks, marketing organizations, research institutes, extension services, etc. The effectiveness of price policies will depend upon whether these other measures have been adopted in the past or still remain to be implemented, and what the sequence of implementation is.

In view of the already mentioned difficulties of generalizing in this area, the discussion will be devoted to some of the main issues relevant to a formulation of pricing policies in relation to the five objectives, and to the typology of countries and crops. We shall see that the typology does not lend itself to a hard and fast application, but should be used more as general guidelines for investigation and analysis. Responses are too fluid and complex to be readily subsumed under single categories, and will tend to vary with a host of other factors.

4 What Are Price Policies?

Prices fulfil three functions: they are signals, incentives, and instruments for the allocation of resources and incomes. Signals are indicators of shortages or gluts, for example, without necessarily carrying with them any incentives to follow them. Incentives are promises of profits for following the signals or threats of losses for disobeying them. Instruments are any means used to achieve the objectives. In principle, these functions can be separated, so that, say, a high price of an item signals scarcity, but resources are not permitted to flow into greater production of that item because an indirect tax is interposed. Or shadow prices can be used for allocating resources, while no financial incentives encourage this. It is possible to imagine an economy in which all prices are zero, all goods are rationed, and all incentives are moral. In a free market economy, prices combine the three functions. The main purpose of price policies is to achieve objectives that, in their absence, would not be achieved.

The existence of price policies suggests the active intervention of government for a purpose or several purposes which are thought to be better achieved by such intervention. But a policy that permits market forces without government intervention to determine prices is also a policy (though not a price policy), if only a policy of governmental self-restraint. As we have seen, in many countries the legacy of numerous past *ad hoc* interventions has left a system that, often without specific intention, discriminates against agriculture, and more specifically against the small agricultural producer. To reverse this trend and give greater scope to price determination through market forces then involves a conscious policy, even though it consists in undoing the 'distortions' caused by past policy. The restoration of neutrality as between the urban and rural sector, between exports and home production, and between industry and agriculture, is frequently a step in the right direction.

If specific interventions are contemplated, the choice is between using prices as instruments of government policies and direct intervention through controls and government activity. But, as we have seen, the government may forswear any form of intervention, whether direct of through prices. Although this is not a price policy, it is a policy of consciously chosen restraint. It is therefore important to distinguish between *laissez-faire*, a policy which permits prices to be determined

11

solely by the free play of market forces, and price policy, which attempts to use prices as instruments of policy, but works through the market. *Laissez-faire* may be inconsistent with free trade, or with competition, which may have to be enforced or achieved by government action. The use of price policies does not imply a commitment to any particular political philosophy and is consistent with any political regime.

Even a *laissez-faire* policy is consistent with forms of government intervention that aim at making the market more efficient. These might apply to the provision of information or of certain goods such as security, storage, roads, training and education, or research, which are likely to be undersupplied if left entirely to private initiative.

Indirect taxes and subsidies, tariffs and exchange rates, as well as anti-trust and anti-monopoly legislation are instruments which governments can wield, while using the market as the mechanism of resource allocation. Governments can also add to demand and supply by operating stocks and thereby achieve a degree of price stabilization or minimum floor prices if they have adequate resources and foresight in predicting the correct future prices. Operation of such buffer stocks or counter-speculative stocks is, however, very difficult, and there is always the twin danger of being burdened with growing excess stocks, or running out of stocks altogether.

Rationing, licensing, prohibitions and government activity not subject to market forces are forms of intervention with the working of the market. In view of the scarcity of administrative talent in many developing countries, and the temptation to corruption resulting from bureaucratic power, there is a presumption in favour of using price policies rather than administrative controls wherever possible. But the transition from the latter to the former hurts vested interests and for this reason is often difficult.

State intervention, in a mixed economy, implies competing or interfering with private agents. If their talents and skills are to be harnessed to the national development effort, it is important to win their cooperation, and to minimize temptations to turn to the black market, to smuggling or other illegal activities, or to lobbying, log-rolling and other pressures on the political and bureaucratic apparatus.

But with all the deficiencies of interventions, developing country governments will wish, and in many cases rightly so, to intervene in the free working of the market for three main reasons. First, they will want to correct what they regard as undesirable consequences in income distribution. Such interventions may have their costs in allocative efficiency, but these costs may be regarded as acceptable. Second, markets in developing countries rarely conform to the tenets of neoclassical economics. In particular, labour costs do not reflect social opportunity costs in the face of unemployment and underemployment. Third, domestic price policy will not always reflect international opportunity costs.

The free market approach has to face the fact that rural labour and capital markets are fragmented, that land markets function poorly, that primary product prices are very unstable and difficult to predict, that information is lacking or inadequate, and that many markets simply do not exist, such as those for future contingencies, which would reduce uncertainty and risks. The government intervention approach, it is now frequently emphasized, is inefficient, leads to 'distortions', is sometimes corrupt, and makes high demands on scarce administrative skills. A better starting point is the use of prices and markets as instruments of policy. This means neither leaving all decisions to the free play of market forces, nor adopting universally direct government intervention and government assumption of economic functions, but in many cases indirect intervention by using prices and markets as means of achieving policy objectives.

Price policy is concerned not only with the *level* of food prices (compared with border prices and the margin for transport, storage and processing) but also with their *structure*, that is the relative level of prices for different crops that may be substitutes or complements, in consumption or production, their *predictability*, since it is future not current prices that matter for supply decisions, their *effectiveness* (i.e. whether the promised price is actually paid by the marketing boards) and their *stability*.[1]

5 Allocational Efficiency and Higher Food Production

A rise in the producer prices of food crops towards or above the level of world prices has two distinct effects: it increases the means and capacity of the grower to produce more food, and it also increases the incentive to do so. There is a will and there is a way. It would be possible to separate the two effects. Lump-sum subsidies with prices unchanged would provide the means without the incentives,[1] while higher prices for particular crops combined with a land tax or a poll tax that takes away what farmers gain from higher prices would provide the incentives but not the means. It is now generally accepted that farmers are responsive to price incentives and that production will tend to increase when rewards are greater. The question is not 'are farmers responsive to prices?', but 'how responsive to what intervening variables other than price, and on what assumptions about other prices?' The response is quite different in magnitude according to whether the price of a specific crop is raised relative to others[2] or whether the prices of all agricultural products are raised relative to the prices of industrial products. Supply responsiveness is much greater in the former case than in the latter, and evidence from one must not be used to support conclusions for the other.[3] When the price of one crop goes up relatively to others, farmers will tend to switch resources from the less remunerative to the more remunerative crop. The response is measured by the cross elasticity of supply, which measures the percentage reduction in some other crop (say cotton) in response to a small percentage increase in the price of a given crop (say wheat).

Lowering the official price of one crop, relative to others, can have the following effects:

1. Farmers may produce less of it and take more leisure.
2. Farmers may switch to other crops, the prices of which have not gone down. As a result of price control of wheat in Mexico there has been a massive switch to sorghum, the price of which is uncontrolled and which is fed to animals, consumed by the better off.

14

3. Farmers may switch to unofficial, sometimes illegal trading channels, where they can earn more.
4. Farmers may produce more for their own consumption, thus contracting out of the market.
5. Farmers may leave the land and seek employment in other sectors, such as migrating to the towns.

There are relatively few studies of total agricultural supply response.[4] This is not surprising in view of the fact that in estimating total agricultural response to a change in the agricultural–industrial terms of trade, other things are never equal, and it is therefore extremely difficult to isolate the impact of relative price changes on total agricultural output. Voltaire once said that you can kill a flock of sheep with incantations – if you add enough arsenic. Similarly, you can get high responses to price increases, if at the same time you improve extension services, add roads to get the produce to the market and improve credit institutions. Whatever results researchers have got on total supply responses to prices only have not been very significant.

Another problem about estimated supply responses is that an increase in the supply of officially marketed output can occur through the reallocation of existing supplies from local (especially rural) markets, with the result that food availability and consumption in the rural areas is reduced. There is evidence from Zambia, Ghana, Kenya and Ivory Coast, that this has happened. It is the reverse of the case where a rise in price leads producers to consume more and sell less of their crops.[5]

Various studies estimate the total short-, medium- and long-term elasticities as of the order of 0.1, 0.2 and 0.4 respectively.[6] Bond's average long-run price elasticity for nine African countries is 0.15. Binswanger *et al.*[7] find short-term aggregate supply elasticities in no case exceeding 0.06, and the highest long-term elasticities in the best cases to be only 0.3. Willis Peterson has argued that the low long-run elasticities found in the time series of country studies were not valid because they do not reflect the impact on investment of maintaining prices high for a long time, so that expectations are changed.[8] Peterson estimated that the long-run aggregate supply elasticity is between 1.25 and 1.66, much higher than other authors' estimates. But this supply function is based on the tacit assumption that the only difference between agricultural supply in developed and developing countries is the structure of price incentives and the volume of research expenditure.[9] The whole burden of the argument of this book is that the absence of non-price measures (availability of credit, fertilizer, and assured water supply, transport, communications, tenurial systems, land distribution and many others), many of them in the public sector, some of them public goods, impedes agricultural responses to prices by themselves in developing countries. In addition, there are differences in soil, ecological and geographical conditions. Peterson's elasticity links illegitimately a series of points on outward-shifting supply curves, each of which curve is much steeper than the curve linking these points.

If we then assume very optimistically that the elasticity is 0.4, that would mean that an annual rise in price of 5 per cent would raise agricultural output annually by 2 per cent. When we say responses of total agricultural output to higher prices are low we are not saying that farmers are stupid, inert, lazy, or subsistence-minded, but that they suffer from severe external constraints in the face of their desire to earn higher incomes.

Supply responses are different for (a) the supply of labour, (b) the supply of agricultural output, (c) the supply of a marketed surplus, and (d) the supply of the officially marketed surplus. Labour supply may go down while crop supply may rise if there is a switch to a less labour-intensive crop, the relative price of which has risen. Agricultural output may rise and marketed surplus may fall, if more food is retained in the household. Total marketed surplus may rise while officially marketed output may fall if farmers are selling more to private traders at higher prices. There is evidence that the total supply response of labour, of agricultural output, of marketed surplus and of officially marketed output will tend to be low or can even be negative, unless surplus land, labour and technologies are available.

The supply responsiveness can be subdivided into four distinct, though related responses:

1. Additional labour use if its supply is abundant, and additional land use for agricultural production. With a few exceptions, the scope for additional land use in the future is limited, and the main response will have to come from the other three responses.
2. Fuller utilization of a given land area with given technology through multiple cropping. Control and allocation of water is often the crucial issue in raising the efficiency of irrigation.
3. Application of new technologies such as high-yielding varieties, fertilizer, water, leading to greater yield of one or several crops per unit of land, per crop.

4. Changing crop pattern to crops of higher value, both by yield and by price.

Underlying these four interrelated causes of supply responses are several more fundamental factors such as the state of agricultural knowledge, institutional arrangements, and technical progress in both agriculture and related sectors such as transport, storage and marketing, the ratio of population to land and the rate of population growth in the rural sector, the constraints imposed by climate and soil, etc.

The supply response can thus be analyzed in three steps. First, there is the response along a given supply curve, with technology, infrastructure, institutions, etc., given; second, a shift of the supply curve that can be attributed to the effect of prices on innovations; and third, a downward shift resulting from non-price-induced measures.

COMMODITY AGREEMENTS

It should be noted that not in all cases is it desirable to increase agricultural production. Where crops are in world surplus, or where an inelastic demand can be used to raise total revenue by price increases and supply restrictions and producers' associations exist of can be brought into being, incentives should be set to *restrict* output, and simultaneously raise receipts and free resources for other uses. The purpose of these incentives is to raise total foreign exchange receipts and to use these to diversify into other activities.

It is, of course, true that the elasticity of demand for a small country's agricultural exports is higher than that for the total world supply of the product, and that particular countries can gain from either defecting from commodity agreements or staying outside. They can hope to benefit from price reductions even though total world demand is inelastic. But such gains tend to be self-defeating if adopted by many countries and the solution should be to reach agreements on the global restriction of producing such commodities. Lenders and donors often take a country-by-country view rather than a global view, and advise country A to diversify into country B's crop, and country B to diversify into A's crop, when both crops are in world surplus.

There are other exceptions to the rule that output should be restricted, and price and total revenue raised where the demand is sufficiently inelastic. A country may be a member of a commodity agreement, but not meet its quota, such as Uganda for coffee. At the agreed price, the demand up to the quota is infinitely elastic. More complicated is the issue of the internal dynamics of a cartel or commodity agreement. Any one country may, by producing more than its quota, receive a larger quota at the next session of the cartel. On the other hand, if enough countries do this, the agreement breaks down. Just as war, according to Clausewitz, is the continuation of politics by other means, so a cartel is the continuation of competition by other means. But it is important to bear in mind that in some cases *reduced* agricultural production of a particular crop, and the use of the additional revenue earned and the resources thus saved for other purposes, is the correct solution.

If higher prices as incentives to greater production are in some cases undesirable, in others they are quite ineffective. This is so for the subsistence crops. Small farmers who consume much of their own output and who seek to assure the security of their livelihood by not participating in markets are largely unaffected by price increases over a wide range. Countries like Bangladesh or India have a large proportion of such farmers who are largely insulated from price policies.

PRICES AND NON-PRICE MEASURES

Even the most ardent advocates of higher prices for agricultural producers[10] would admit that many other things are necessary in order to call forth the extra supply. Many developing countries have not only kept agricultural prices low, but have also invested a smaller proportion of their GNP in agricultural research, have kept public investment in agricultural infrastructure (e.g. irrigation and transport) low, have starved agriculture of credit and investment in human capital, such as education and health. Since, historically, both price and non-price measures have tended to discriminate against agriculture, the agreement that other things are necessary does not amount to much. Among these other things is the removal of physical, social, and administrative barriers to increased supply. To illustrate: transport bottlenecks must often be broken so that the extra food can be transported from surplus to deficit areas or to ports for export; consumer goods must be available in the countryside, so that the farmer can spend his higher income on something he wants; agricultural inputs must be available so that the technical conditions for raising supplies can be met; the marketing authorities or the private middlemen must pay promptly for the crops bought. There must be institutional arrangements that ensure that the benefits of the higher prices accrue to the farmers and not to monopolistic private middlemen, or absentee landlords, or disliked foreigners, or inefficient or corrupt public marketing authorities.

Above all, new cost-reducing technology must be available, so that the incentives of higher prices can speed up the growth of production significantly. This would not necessarily be the case with traditional technology which would run into rapidly diminishing returns.

The technology is particularly important, for it presents the solution to the dilemma between higher producer and lower consumer prices. For an improved technology can raise producers' returns and incomes, without raising prices to consumers. Such productivity improvements, which lower the short-term supply curve, are the ultimate answer to the dilemma.[11]

Beyond the proposition that a host of other things is necessary agreement ceases. The price advocates argue that many of these constraints are removed as a result of higher prices (or, a point we shall come to later, of lower input prices). Tube wells, land levelling and irrigation channels will be responsive to higher rewards. Higher prices also (the advocates argue) induce the adoption of new techniques and the extension of research. Even public investment and institutions are said to respond to higher output prices. Others are less hopeful of all these blessings following automatically on higher prices and would argue that these technical constraints have to be removed by independent measures. If these measures are taken, the role of prices is changed, in ways discussed later; if they are not taken, prices will not do their work in increasing supply, though they will, of course, give at least higher cash receipts to farmers.

In South Korea, for example, rice yield increased by about 50 per cent between 1970 and 1978. Although there was a fivefold increase in the government farm purchase price of rice, a number of other measures were also taken, such as irrigation which increased from 1.02 million hectares to 1.13 million hectares, and the rapid introduction of high yielding varieties (HYVs) was possible as a result of investment in research. The area under HYVs increased from 15 per cent to 70 per cent between 1974 and 1978.

On the other hand, the adoption of HYVs in Egypt was less successful, in spite of the seed price subsidy rising to 30 per cent and an extension campaign to encourage their use. Less than 1 per cent of the area under rice is under HYVs, compared with 64 per cent in the

Philippines and 30 per cent in India. The grain quality was not acceptable to consumers, the ratio of straw to grain was too low to meet animal fodder requirements, local research was insufficient to make the HYVs as disease-resistant as the local varieties, and the cultural demands of HYVs did not fit into the common farming system.

In Bangladesh, the government's strategy has been to encourage the use of modern inputs such as fertilizer, irrigation and HYVs, and to provide production incentives to farmers. Yet, agricultural growth has been slow towards Bangladesh's target of achieving grain self-sufficiency. Since the late 1970s public investment in agriculture was expanded and a series of devaluations may have raised price incentives to farmers. Irrigation, as well as improved incentives, have contributed to growth in agriculture. But there has probably been an increase in landlessness and rural unemployment. Recently, food production has kept up with population growth, marketing has been handed over to private traders and subsidies have been reduced. Some observers have hailed policy reforms as a great success, but weaknesses remain.

A study of Bangladesh shows weak responses to price increases by themselves, but if accompanied by improved technology, transport and storage facilities, more irrigation and better water control and management, and other measures, there is a strong response to price.[12] Further progress will therefore depend upon (a) the proportion of actual and potential surplus farmers are producing for a market, (b) the removal of physical and social constraints, (c) the availability of, and ability to adopt new technologies and (d) rising levels of demand that sustain the price incentives.

The fundamental difference between price protagonists and price sceptics is on the question whether raising prices without these other measures is better than nothing, or whether, by itself, it is futile or can even be counter-productive. The critics can point to cases where the introduction of higher supply prices, without an appropriate scale-neutral technology, without the appropriate education and without the appropriate infrastructure and institutions has, for example, accelerated the transfer of land from small to large farmers and violated equity and anti-poverty objectives. If the productivity per acre on the small farms was greater than on the large farms, total output falls. This is likely to be the case because small farmers apply more family labour to the land. In such cases, higher prices can lead to reduced food production, aggravated inequality, and greater poverty.

The proposition 'other measures, in addition to price policies, are necessary', to which we gained initial general consent, is therefore open to two diametrically opposed interpretations. It may mean that while other measures help, getting prices right without them is better than nothing; or it may mean that the right prices together with the institutional and technological measures will achieve the objective, but by themselves will be ineffective or may be counter-productive. If the position that other, non-price, measures are necessary in order to make price policies effective, the difficult question arises about the correct sequencing, since it is highly improbable that all measures, price and non-price, can be instituted at the same time. If the non-price measures, such as research, technology, irrigation, transport, etc. are already in existence, the introduction of incentive prices can be expected to do its work. But if the non-price measures are absent, incentive prices will tend to be futile or may even achieve the opposite of the intended objectives.

Various studies have shown the nature of the appropriate phasing of price and non-price measures. K. N. Raj argues that the high rates of growth of agricultural production in the People's Republic of China since 1970, often believed to be the result of more favourable price policies, can be attributed to investment in infrastructure, water, chemical fertilizers and soil conservation in the 1960s.[13] Hayami and Ishikawa have pointed to the importance of irrigation in Japan and Taiwan, and the slower growth of agriculture in the early stages in Korea because of lags in investment in infrastructure.[14] Tibor Scitovsky has shown[15] that in South Korea and Taiwan – two success stories in agriculture, but not only in agriculture – research stations, a network of extension offices, the provision of inputs such as seed and fertilizer in kind, lending of equipment, organization of cooperatives both for marketing and for the distribution of credit and fertilizer, and the construction of an infrastructure of roads, railways and harbours had been instituted by the Japanese during the colonial period, especially in Taiwan, where the climate was more favourable, and where colonial rule lasted longer (fifty years) and rulers and ruled got along better than in Korea. In addition, there was a radical land reform. These are precisely the conditions in which price policies and the market can be applied with the best effects. On the other hand, there is a presumption that in Argentina, where some of the non-price measures are 'in place', low agricultural prices due to taxes on agricultural exports and overvalued exchange rates, have discouraged agricultural production.

It is possible to take four different views of the role of non-price measures (e.g. research, technology, education, health, extension, credit, transport). Either it could be the case that they are induced by price measures and no independent action is necessary. Inducement may even apply to the public sector. Hayami and Ruttan have shown that some innovations have responded to the relative factor endowments of different regions, and that land-scarce, labour-abundant societies have more biological–chemical innovations, while labour scarce, land-abundant societies have more mechanical innovations.[16] They have also argued that relative factor prices affected not only technological development but also the design of social institutions. One important factor affecting the extent of the 'inducement' are the budgetary costs involved. If government expenditure on innovation and infrastructure is not constrained, the induced impact will be much larger than if it is. Once again, we observe the crucial importance of the budgetary constraint. It is clear that prices can influence the profitability of different techniques made available through applied research, and therefore their rate of adoption and diffusion. But by themselves they cannot explain the growth and direction of basic scientific knowledge and of public investment in research, extension, infrastructure and human capital. The growth of basic scientific knowledge cannot be explained in terms of some linear function of prices.

Or, secondly, it could be that the non-price measures are already in place, so that prices can 'bite'. This is the case in most advanced industrial countries, in which agricultural markets are also more commercialized and in which the share of non-agricultural inputs is high, the supply elasticity of which is large. It is mainly for this reason that data on supply responses to price increases collected in advanced countries cannot be directly applied to low-income countries. Or, thirdly, it could be that only in conjunction with non-price measures are price measures effective, and they require independent action. Or, fourthly, it could be that they are substitutes for price measures, so that prices need not be raised at all, or by less, if they are taken. Non-price measures can be substitutes for price measures if direct taxation of agriculture is ruled out, and the only way to finance public investment in agriculture is to squeeze the agricultural–industrial terms of trade. This is not a justification for transferring resources out of agriculture, but for using, in certain circumstances, price measures as taxes to finance non-price measures that yield a higher response. In the third case, price measures without non-price measures are ineffective or counter-productive, in the fourth case, they are unnecessary. Experience shows that each of these is a possibility in different circumstances.

For example, in Malawi smallholders are performing well in spite of receiving only 20 per cent of the auction price for tobacco or 60 per cent of the export price for groundnuts because the supporting non-price services are efficient and reliable.[17] In Togo cocoa production declined, in spite of higher real prices, largely because of inefficient sharecropping arrangements, while cotton production grew dramatically, in spite of prices remaining the same.[18] Raj Krishna has shown that, while ideally technology policy should go together with price policy, a good technology policy is more important in stimulating output.[19]

Tanzania presents a good example of the futility of price incentives by themselves. Lack of foreign exchange reduces industrial capacity utilization to less than 25 per cent. An increase in exports is essential. Increased inputs into agriculture and transport, both road transport and port facilities, and more consumer goods for farmers to buy are a precondition for getting exports up.[20] The IMF requires a large devaluation, the removal of price and wage controls and import liberalization. But without the improvement in agricultural inputs and transport, and the availability of consumer goods, inflationary changes in prices resulting from devaluation would have little effect or can be counter-productive and disruptive. There would not be an increase in the incentives to farmers, and the consequential price inflation may even reduce supplies. With these non-price changes, a much smaller devaluation may be needed, if only to improve the financial position of the marketing boards which are running large deficits.[21]

Many of the non-price measures call for action by the government. The best price incentives are of no use if there are no roads to carry the crops to the market. Supply cannot respond to the promise of higher profits in the absence of credit institutions that provide the farmer with money to buy fertilizer, pesticides and equipment. Without adequate irrigation and water control the supply curve may rise steeply. Above all, the seed–water–fertilizer technology must be available which is itself the result of research, directed at profitable smallholder cultivation of crops. Much research can be carried out only in the public sector, because private agents could not capture the whole or even a substantial part of the returns, and because the minimum economical size is too large for small farmers. Basic research in genetics, plant physiology, soil science, etc. is of this nature. In addition, much research into new plant varieties is expensive and risky, but the result can easily be imitated. Hence it does not pay farmers or firms to conduct such research.

Several studies of Bangladesh (Bertocci, Geoffrey Wood, Swedish

International Development Agency)[22] and of Indonesia (Rex Mortimer)[23] report increasing land concentration and growing landlessness as obstacles to rural progress. In Bangladesh, in 1977, 3 per cent of the households owned more than 25 per cent of all land, and 11 per cent owned more than 52 per cent. One third of rural households own no agricultural land and, together with those owning less than half an acre, the virtually landless constitute about 48 per cent of the rural population. In such a situation an increase in the price of food, which raises the wages of rural labour, may lead to a switch to less labour-intensive crops, e.g. from jute to rice, and to a decline in the demand for rural labour.

As David Felix has pointed out, the expression 'getting prices right' has undergone a curious transformation.[24] In the 1960s it was meant to point to the calculation of correct shadow or accounting prices in the face of 'distorted' market or actual prices. Since market prices reflected all sorts of 'distortions',[25] including those caused by the existing and from an ethical point of view arbitrary income distribution, it was the task of government to intervene and allocate resources according to the 'correct' shadow prices. It was the purpose of government intervention to correct the distortions produced by the free play of market forces. Ten years later, the recommendation was reversed. It was then assumed that developing countries should get rid of state interventions so that their market prices would reflect international or border prices, which reflected the 'correct' opportunity costs. This can be true only if it is assumed that each country is so small as to be incapable of influencing its terms of trade. Altogether different principles apply to non-tradable goods and services. Considerations of income distribution also went by the board. True, the existence of market failure does not *ipso facto* justify government intervention. But equally, 'government failure' does not *ipso facto* justify reliance on the market.

Once it is agreed that 'getting prices right' is an important element in a strategy, the question arises what are the 'right' prices? In deciding what guidelines to use in determining agricultural prices, the prime candidate is international prices. These represent the opportunity costs of tradables (at least for a small country, incapable of influencing its terms of foreign trade). World prices represent the potential earnings if the crop were exported, or the potential foreign exchange costs, if imported. They should therefore be used as a benchmark, allowing, of course, for the costs of transport, storing, marketing and processing incurred between the local producer and the price at the port. But some departures from this benchmark are justified. The most important departure is the need to prevent prices from fluctuating with short-term movements in international prices. Wide fluctuations can convey the wrong signals and incentives because agricultural responses take time, and this lag may aggravate fluctuations in supply and price. Moreover, the uncertainties caused by fluctuating prices have detrimental effects on the investment decisions of low-income farmers who cannot afford to take risks because shortfalls would bring them below the subsistence minimum. Agricultural investment and innovation are then depressed.

Some countries export food crops in years of good harvests and import them when harvests are poor. The import price (c.i.f.) is the relevant benchmark if the food is imported, the export price (f.o.b.) if it were to reduce potential exports. If a country imports food in some years, and exports the same food in others, as for example maize and sugar in Kenya between 1979 and 1983,[26] the price is indeterminate within the f.o.b.–c.i.f. range.

The fact that most advanced industrial countries protect their agriculture makes international prices lower than they would be in a free competitive market. I say 'probably' because some forms of agricultural protection are so inefficient that agricultural production migh conceivably be greater, and world prices lower, in their absence. The US Government has for example paid farmers for *not* growing wheat. If world prices are unduly depressed by rich country protectionism, this by itself does not detract from using these prices as a benchmark. Only if there is a possibility that the protection will cease and future prices be higher, or if protection leads to greater price fluctuations, is there a case for departing from the benchmark. The world price may also be the price of a relatively small residual market that provides the wrong signals for domestic pricing policy, as at some periods, when there are major international commodity agreements, in the case of coffee or sugar.

Moreover, in a world of fluctuating exchange rates the determination of what are international prices is uncertain. Today, international capital movements are perhaps a hundred times the movements of trade, and it is these that determine exchange rates. The determination by international prices of what are comparative advantages has got lost in the sea of capital and currency transactions. What foreign exchange rate should be used to convert the international price into a domestic price quivalent? If a developing country largely trades with one major industrial country, the answer might be thought to be easy. But it may export to one area and import from another. Clearly, for investment decisions in agriculture anticipated future prices are relevant, and current world prices are not always a clue to these, though futures markets, where they exist, may be.

The EEC protects some agricultural products, such as dairy products and sugar, excessively. It would constitute an argument for raising domestic producer prices in developing countries above the level of world prices, as an insurance against the risk of reduced world supply. It is therefore an added reason for raising prices. But a dilemma can arise: agricultural prices may be too low in relation to non-agricultural domestic prices, and at the same time too high in relation to international prices.

A good rough guideline is: keep domestic cereal prices in line with an estimated trend of future world prices (estimated by a reputable authority), calculated at some major convertible currency, or at the currency of the country's major trading partner, plus some percentage increase for importables to allow for uncertainties in world supply, unexpected rises in world prices or unavailabilities.

The EEC's variable levy on agricultural imports achieves a degree of price stabilization within Europe. It will be argued in a later section that this is, in principle, a desirable objective, particularly from a macroeconomic point of view. But the EEC levy does this by destabilizing world prices. In times of global shortage and high prices the levy goes down, adding to world demand, and in times of glut the levy goes up, subtracting from demand. Demand is thereby made less elastic. The lowering of the levy also reduces supply in times of shortage, while its rise increases it in times of surplus. Supply is therefore also made less elastic. The reduction in the elasticities of demand and supply is likely to make for greater instability of world prices in the face of random shocks.

The case for raising producer prices of agricultural products is, as we have seen, based on the very widespread practice in developing countries of paying too low prices to producers. The methods of doing this include domestic price fixing, implicit and explicit taxes, either in the form of export taxes or through price controls often accompanied by compulsory government procurement or sales to parastatal marketing authorities, or other forms of government monopoly in domestic or foreign food trade. Another common method is overvalued exchange rates which represent an implicit tax on agricultural exports and a subsidy to agricultural imports competing with domestic production. Tanzania, for example, has paid farmers throughout the country a low price for maize, encouraging the sale in black markets. Thailand has traditionally taxed rice exports with detrimental effects on production. In Mali, a study has shown that it costs farmers eighty-three Malian francs to produce a kilo of rice but that the government paid farmers only sixty Malian francs per kilo.[27] A World Bank analysis showed that for major crops in thirteen African countries between 1971 and 1980, taking the net tax burden and the effect of overvalued currency into account, producers in these thirteen countries received less than half the real value of their export crops.[28]

In some cases it is possible to raise prices to producers without raising them to consumers, or to lower them to consumers without lowering them to producers. This is the case where *marketing margins* can be reduced by raising the efficiency of marketing and lowering the costs of transport, storage, processing, etc. This will normally involve investments in improving roads, facilities for distributing fertilizers, milling facilities, etc. In Indonesia, marketing margins for rice have narrowed from 30 to 40 per cent in the mid-fifties to about 10 per cent in 1979.[29] In many other countries such as Tanzania, Mali, Zambia and Kenya there is scope for narrowing margins, thereby raising incentives to producers without harming consumers. There are cases of successful cooperatives that have taken over the services of middlemen, such as marketing, transporting, storing, processing and thereby not only appropriated their profits but also improved the efficiency. This kind of 'import substitution' at the local level, i.e. incorporating transactions previously carried out by agents outside the community, lies at the root of many successful agricultural producers' cooperatives. Examples are Kurien's dairy cooperative in India or the Bombay fishermen's cooperative.

The origins of many state marketing and trading companies such as CONASUPO in Mexico lie in the desire to protect small farmers from being exploited by private monopolistic traders. But many of the parastatal marketing boards have now drawn a good deal of criticism for being inefficient, corrupt, absorbing excessively high margins, or delaying payments to farmers. The call has gone out to 'privatize' them and substitute private for state trading. But it is clear that efficient and inefficient or monopolistic practices can take place in both the private and the public sector. An investigation in Tanzania, in 1981, showed that the margin going to the grower was much the same in sisal, which is marketed by a parastatal, as in bananas, which are privately marketed. A similar situation was found to be true for coconut oil in the Philippines, where a private multinational corporation could set a low transfer price for the product exported to its US parent. The problem is to establish efficient marketing authorities, whether private or public. In some Latin American and African countries, private traders are permitted to compete legally with parastal government agencies. This can contribute to greater efficiency, or it can lead to a form of accommodation, so that margins are set so high that the least efficient private trader can survive. Of course, measuring margins in marketing and their change over time provides no evidence for whether they are too high or too low. Everything depends on the costs of the services rendered by these authorities, such as transport, storing, processing, etc. These costs can be high and rising, e.g. with the cost of energy.

A study of marketing in Kenya collected

> detailed information on the knowledge of prices held by different traders, agreements of collusion between traders, costs of bribing the police, and the approach to formulation of prices. Among other things it was found that restriction of maize movement tended to discriminate against small traders and that rural producers, consumers, and the urban poor suffered most as a result of marketing inefficiencies.[30]

In order to get a fuller picture of the incentive system one has to look not only at output prices but also at the prices of inputs and the consumer goods bought by farmers. The supply elasticity of agricultural crops depends also on the share of inputs bought from the non-farm sector, and on their elasticity of supply. Since their supply elasticity is often large, the larger the share of these inputs, the higher will be the supply elasticity of agricultural products. If, on the other hand, the share of non-farm inputs is small, the fact that their supply elasticity is high will not affect the supply of agricultural products very much.

6 Subsidies to Inputs

Low prices of agricultural products are often partly offset by subsidies to inputs. The most commonly subsidized items are fertilizers, credit, tractors, pesticides, seeds and the services of infrastructure. In some cases the subsidies are intended to compensate for the protection of high-cost industries, such as fertilizer, farm chemicals and tractors. They are then 'distortions' that compensate for other 'distortions'. These subsidies to inputs can be useful to encourage farmers to use a new input, or where external economies are important, so that benefits accrue to others, in addition to the farmer using the input. But they also often have some undesirable and unintended effects. They tend to encourage the inefficient use of inputs, for example, the waste of water. They encourage the use of certain types of input, such as fertilizer, but not of the most abundant factor, labour. Subsidies should never encourage substitution of a scarcer factor for a more abundant factor. Though not inherent in subsidies, as is shown by Korea, in fact they often benefit only or mainly the large and rich farmer and, if accompanied by rationing, can actually deprive the small and poor farmer, who uses his land more productively, of resources. In Kenya, for example, 80 per cent of subsidized fertilizer has gone to larger farmers. If the subsidized input is exported at a profit, or smuggled abroad, the price paid by the domestic producer can be higher than it would have been in the absence of the subsidy. Or it may cease to be available altogether. Bangladesh reduced its initial large fertilizer subsidy after 1979 from 10 per cent of the development budget (and 50 per cent of unit cost) to 2.4 per cent (and 17 per cent of cost). This, combined with large increases in irrigation and water control investment, resulted in *more* fertilizer being available to farmers, and a growth in fertilizer sales of over 10 per cent per year, whereas before there were frequent shortages and high unofficial prices. If different crops, such as cotton, wheat and rice use the subsidised input, say water, in different proportions, the subsidy will encourage increased production of the crop using most of the input, at the expense of the others, again often an unintended result.

Credit subsidies have been criticized on the ground that they failed to achieve the objective of helping small and poor farmers. They appear not to have raised agricultural output, nor improved rural income distribution, nor raised rural savings. Credit institutions have been weak, inefficient and sometimes corrupt. There are, however, some success stories. These have depended on the design of efficient credit institutions.

The ideal system of combining low consumer prices with high producer prices is a system of deficiency payments: farmers receive subsidies according to the value of their output. It worked well in Great Britain before her entry into the Common Market, but it is expensive and administratively cumbersome. Its budgetary burden is beyond what most developing countries can bear.

7 The Total Picture

Even taking producer and input prices together does not give us the whole incentive picture. Low producer prices are aggravated by open or concealed taxes or tariffs on the consumer goods bought by farmers. The total twist in the income terms of trade between agricultural and industrial producers is the result of depressed supply prices, high prices of consumer goods (mainly as a result of protection) and some selective subsidies to input prices.

In some countries such as Tanzania the prices of consumer goods are controlled, but they are rationed and often not available in the villages for farmers to buy. If consumer goods are not available to the farmers, either because they are rationed, their prices are controlled, and black markets are successfully suppressed, or because of transport or other bottlenecks, higher prices for the cash crops will lead to reduced supply. A smaller quantity of crops sold will then be sufficient to buy the limited quantity of consumer goods. The elasticity of supply will be minus one. On the other hand, without any increase in price, supply will rise as more consumer goods become available. Only if the future availability of consumer goods is uncertain, and if farmers wish to store money in anticipation of availability, may this perverse effect be reduced or avoided.[1] But if farmers store money in anticipation of future availability, when these goods become available supply may be reduced. Paul Collier also envisages a cumulative contraction of agricultural supply. Assume (not unrealistically) that urban groups have a claim to a certain fixed amount of imports and that the cash crop is exported to pay for these imports. If then farmers cannot buy the desired quantity of imports out of their incomes, they will contract supply, which will reduce imports, which, with the constant urban claims, will further reduce availability in the countryside, which will reduce crop supplies, in a cumulative contraction. In Collier's model, the amount produced for consumption by the family (or leisure) increases as sales for the market contract. This model assumes that the government is successful in eliminating all black and parallel markets[2] and in keeping the prices of consumer goods down. For otherwise, the rationing will tend to drive up the prices in the countryside and not reduce farmers' supply.

A survey of policies in fifty developing countries by the World Bank found that forty-six provided disincentives to agricultural production.

The measure of the disincentive effect of these policies is the effective rate of protection received by agriculture compared with industry. Studies of eight countries showed nominal and effective rates of protection that permitted farmers to receive only 50 to 80 per cent of the farmgate equivalent of export prices for their major crops. In contrast, industrial goods enjoyed levels of protection from 30 per cent to 200 per cent above world prices. A World Bank study on nominal and effective rates of protection for twenty-one countries in the seventies showed that effective protection rates for food grains clustered around 0.80, the greatest divergencies being Chad (with coefficients of 0.44 and 0.52 for rice) and Korea, Ecuador and Yugoslavia (1.15 to 1.29 for selected cereals). Taking into account subsidies to inputs such as fertilizer and credit, the total distortion in the twenty-one countries is equivalent to a tax rate of between 30 and 40 per cent on growing food.

Another study[3] showed that in the seventies, agriculture was heavily discriminated against in Brazil, Cameroon, Ghana, Nigeria, Senegal, Tanzania and Uruguay, and highly protected in Bolivia, Korea and Peru. This study also finds countries with high levels of discrimination against agriculture had lower agricultural growth rates and lower export growth rates than those that discriminated less, though there was no difference in aggregate growth.

The net effects of these distortions often include the following undesirable results.

1. Discouragement of agricultural production, and reduction in rural living standards.
2. Discrimination against the small farmer, who has to sell his surplus at harvest time at a low price, because he has no storage capacity and no access to cheap credit (e.g. India as a result of the state procurement system).
3. Encouragement of evasion of controls, black markets, smuggling and corruption.
4. Where the prices of some food items are controlled while others are free, there is encouragement to a switch to the production of the uncontrolled, such as fruits, vegetables and livestock, mainly consumed by the better off.
5. Encouragement of capital-intensive techniques (if e.g. there are subsidies to tractors).
6. Encouragement to new industrial activities that diverge from the optimal.

7. Discouragement of exports and foreign exchange earnings.
8. Encouragement of imports, sometimes dependence on food aid, frequently of luxury crops that change consumers' tastes in favour of imported grains.
9. Reduction in employment.
10. Increased rural–urban migration.
11. Negative value added in processing (e.g. Ivory Coast 1967–72).
12. Encouragement of imported equipment.
13. The creation of industrial monopolies.

The removal of these distortions would require a multipronged attack on the following six fronts:

1. Incentives (or prices);
2. Inputs;
3. Innovation (technology, including irrigation);
4. Information (the diffusion of the technology through extension);
5. Infrastructure;
6. Institutions (credit institutions, marketing, land reform).

(These are the six 'Ins' or *Ins*truments that provide a mnemotechnic framework for agricultural policy.) Some of these can only, or more efficiently, be provided by the public sector.

We might add to these as the seventh 'I', Ignorance.[4] (Or, if we wish to stick to 'In', Innocence.) It is to remind us that we do not know all the answers, and should keep an open mind to learn from our mistakes. As Andrew Kamarck said, inconsistency (the eighth 'In') can be a virtue in the face of ignorance. It can be a way of learning.

Clearly, complex and difficult choices arise within each of the six 'Ins'. Consider infrastructure, normally provided at least partly at public expense. There are choices between physical, legal, human, social and producer-specific types of infrastructure to be made; choices between centralized and decentralized types of infrastructure, choices between infrastructure for producers and for consumers, choices between expenditure on maintenance and new projects, and choices between different methods of financing expenditure on infrastructure. The same is true for institutions and for information. (For example, should extension workers concentrate on single lines of conveying or should they combine several?) All these compete for scarce resources with directly productive investment in agriculture.

The argument has been that price incentives work only in conjunction with actions on the five other fronts, and that private supply responses depend on public policies. But this does not mean that many governments have not spent excessive amounts of money on some types of central infrastructure, particularly airports and four-lane highways that have mainly benefited urban groups and rich farmers. The plea for public sector action must not be misinterpreted as a licence for reinforcing extravagant urban bias.

As we have argued in the context of the use of international prices, there is a case for going beyond the removal of distortions as measured by international prices and giving additional incentives to food production, because private risks are often greater than social risks, and, through the adoption of innovations and learning effects, the ultimate cost reduction can more than compensate for a temporary excess level of incentives. It is essentially an application of the infant industry argument to an agricultural sector.

In the efforts to raise producer prices as an incentive to agricultural production, the benefits can be undone if the higher prices are communicated to the rest of the system through proportionally higher money wages and higher mark-ups on industrial goods. If this were to happen, the attempt to raise agricultural prices would only set off an inflationary movement of the general price level, would not have improved agricultural–industrial terms of trade, would not achieve its objective of raising food production, and would have undesirable side-effects.

8 Aggregate Growth

The second objective of food price policies is to encourage industrialization and accelerate aggregate growth. It used to be thought that,
quite apart from equity considerations, low food prices will tend to
keep industrial wages low and thereby contribute to faster total
growth.

The links between agriculture and industry were formulated in the
now discredited doctrine of the surpluses,[1] though remnants of this
doctrine still linger in many policy-makers' minds. This doctrine
advocated the need to squeeze out of agriculture five surpluses: a
labour surplus, an investable surplus, a marketable surplus, a foreign
exchange surplus, and a budgetary surplus. Since direct taxation of
agriculture is regarded often difficult or impossible, the surpluses are
to be mobilized by twisting the terms of trade against agriculture.

It is now obvious that industry, far from needing a *labour surplus*,
cannot absorb a fraction of the new entrants into the labour force and
that employment opportunities have to be found in the rural sector:[2]
not only by efficient labour-intensive techniques in agriculture, but
also by rural public works, rural industries and rural services, that can
use agricultural labour when seasonal demand is low. The combination
of the initially small industrial base in low-income countries, of labour-
saving industrial technology, and of the rapid increase in the labour
force have changed the situation radically from what it was in
eighteenth and early nineteenth-century England, on which early
policy recommendations were based. The present system in many
developing countries of low food prices and depressed rural incomes
drives rural people into the towns far beyond the number needed by
industry, and adds to unemployment, underemployment, low industrial wages, and the high costs of urban services.

The doctrine of the *investable surplus* was based partly on historical
analogies and partly on two assumptions. Historically, the now
developed countries, and most recently Japan, have drawn on
agricultural savings for the investment in non-agricultural sectors. In
addition, the doctrine was based on the belief that agriculture had (a) a
low propensity to save and (b) a low 'absorptive capacity' for
investable funds. Research showed, however, that savings propensities in agriculture are at least as high as in industry (and among small
farmers nearly as high as among large farmers) and that there are

ample investment opportunities in agriculture and the rural sector generally, which can show higher rates of return than many industrial projects. It is therefore neither necessary to twist the agricultural terms of trade as a substitute for voluntary savings, nor desirable to deprive agriculture of investable resources. For some purposes the distinction between the two sectors is misleading, and the investment should go into agro-industrial projects, which cut across the dividing line.

When rapid industrial growth has been achieved by forcing resources from agriculture into urban industry, including industries that process local materials, the costs in terms of forgone agricultural growth, increases in poverty and inequality, unbalanced geographical development, urban congestion and ultimately delayed industrialization have been high.

A *marketable surplus* of food and agricultural raw materials is, of course, essential if the farmers are to buy industrial inputs and consumer goods (and to repay cash debts), and if people producing non-agricultural goods are to be fed. It has been singled out as the principal limiting factor on development. It also makes a contribution to employment creation in the industrial sector. But in conditions of malnutrition and starvation in the countryside, and, paradoxically, even among those producing food, there is a strong case for policies that encourage the retention of a substantial part of the food grown within the agricultural sector, channelling it from surplus to deficit regions, from rich farmers to poor farmers and labourers, and from the harvest season to the rest of the year. In this light, the possibility of the reduced marketable surplus sold by small surplus farmers to the towns as a result of a price rise for their crops is entirely desirable, if more of their crops can be retained and consumed by them and their families.[3]

This clearly does not mean that any reduction in the marketed surplus raises consumption on the farm. What is argued elsewhere about the conflict between national self-sufficiency and self-reliance also applies to the household. A decline in marketed surplus may also mean reduced production either because more leisure is preferred or because risks due to weather are greater, or it may mean a switch from official to unofficial markets. On the other hand, an increase in the marketed surplus following price rises can lead to reduced local supplies of food and reduced consumption on the farm.[4]

The *foreign exchange* surplus is another contribution that agriculture can make, though here again, it would be wrong if it were achieved at the expense of food for the undernourished people in the agricultural sector. The dilemma between agricultural production for export and food crops for domestic consumption raises complex issues, discussed briefly below. The *budgetary surplus* – an excess of direct agricultural taxation over revenue spent on agriculture – is rare for food crops, though common for export crops. A progressive land tax would meet some of the objectives listed on page 5 with a budgetary gain rather than loss.

It is now commonly agreed that it is undesirable to squeeze agriculture by turning the terms of trade against it, and more particularly against poor rural producers: undesirable not only on grounds of equity, but also on grounds of promoting industry. For a large country, a necessary condition for industrial growth is a prosperous agricultural market. Foreign trade apart, agriculture also provides industry with its food and raw materials, and industry provides agriculture with industrial inputs. The dispute as to agricultural *versus* industrial priorities is largely a sham dispute. It is a case of Both–And, not Either–Or. Agriculture and manufacturing industry are interdependent. The interdependence is both via inputs and via markets and calls for a simultaneous advance on both fronts. A prosperous and efficient agriculture provides the raw materials, the food, and the markets for industry, and a growing industrial sector provides the tools and fertilizer for agricultural production and the markets for its marketable surplus. Some of the complaints against the squeezing of agriculture should be directed against the high-cost, inefficient industrialization for which the resources were used. Had they been used efficiently, they would have benefited agriculture.

Considering only African countries where agriculture has lagged behind population growth, we find that countries that have shown high growth rates of agriculture between 1970 and 1980, such as Liberia, Kenya, Malawi and Ivory Coast, have also registered high manufacturing growth. Countries with negative or zero agricultural growth rates in the same period, such as Mozambique, Ghana and Zimbabwe, also show negative industrial growth. And countries in which agricultural growth was slow, such as Chad, Ethiopia, Upper Volta and Sierra Leone, also show slow manufacturing growth.[5]

While the role of agriculture as a source of these various surpluses – labour, savings, food and raw materials, foreign exchange and public revenue – must be qualified in the ways indicated, agriculture's function as an important market for the output of industry remains, and is increased if the qualifications are taken into account. For the excessive squeezing out of surpluses is in conflict with the creation of a prosperous agricultural market for industrial goods. As the market grows and prospers, the 'marketable surpluses' which are exchanged for the industrial goods, of course, also grow.

At the same time it must be remembered (it not always is) that, while raising the prices of food and other agricultural goods, and improving incentives to grow them, will tend to increase their supply, it necessarily means lowering the (relative) prices of some other goods and, by the same token, blunting incentives for, and reducing *their* supply. In so far as removing 'distortions' improves the allocation of resources, total national production can rise. But the *relative* encouragement of one sector necessarily spells the *relative* discouragement of some other sector. Advocates of higher food prices should always bear this in mind.

9 National Food Security and Food Aid[1]

Food security has been defined in a number of different ways. In its widest sense it means assured physical and economic access to food, at all times, to all citizens. In this sense, the subject covers all aspects of food supply and demand discussed in this book. In many countries, particularly large ones such as India, the main causes of food insecurity are domestic. They comprise low levels and unequal distribution of food, an unequal distribution of land and other assets to produce food, and low levels of investment to increase and to stabilize food production. In other, mostly smaller, countries food insecurity arises from large changes in the price of imported food or possibly from the threat of political food embargoes. Various authors distinguish between chronic and transitory, anticipated and unanticipated, regular and random, and seasonal and year to year food insecurity. For the purpose of this brief section, we confine the meaning to reducing fluctuations in food consumption and to the international contribution to food security. Special attention will be paid to the role of food aid as a contribution to national food security.

Self-reliance and self-sufficiency, though often identified with one another, are not only distinct objectives but can be in conflict. A country aiming at self-sufficiency in food could become highly vulnerable and dependent on foreign support if its domestic harvest failed, or if it suffered a series of droughts, so that it would have bought attempted self-sufficiency at the cost of independence and self-reliance. On the other hand, a country with a diversified pattern of production and foreign trade, i.e. diversified by sources of supply of imports, by markets for its exports, and by commodities and services, and a country with ample foreign exchange reserves and access to lines of credit, would have achieved self-reliance without being self-sufficient. Or, to take another example, a country depending entirely for fuel on domestic coal is at the mercy of its coal miners and is less self-reliant than one that draws on diversified sources of energy, including some imports. Dependence, for which self-reliance is the remedy, is not constituted by the existence of foreign trade relations, but by trade that is highly concentrated by sources, markets and commodities, so that risks are high if there is a change in policy or

39

demand or technology, and the fate of the country depends on events in one or a few places outside the country, or on one industry or on one type of market.

But while foreign trade is consistent with self-reliance (interdependence, which is consistent with self-reliance, is not the same as unilateral dependence), any country relying entirely on foreign trade for food imports also exposes itself to serious risks. These are not only technical and economic risks but also political risks. To reduce these risks, or to ensure itself against them, a country has several options. It may permanently grow more food than would be strictly indicated by current world prices. This would require raising food prices above international levels, the extra costs representing an insurance premium. Or it can hold reserves of arable land, to be cultivated when the need arises. Or it can increase the areas under irrigation, thus reducing fluctuations due to variations in rainfall. Or it can hold national buffer stocks. These are expensive. (In the Sahel using a ton of grain from a buffer stock is estimated to cost $500 compared with world c.i.f. prices of $200; others have estimated the cost of buffer stocks at $50 per ton per year. Costs depend on the length of time over which stocks are held.) They involve wastage, but reduce the dependence on the vagaries of international trade and shipping. Or it can hold foreign exchange reserves, less costly to store and more flexible. Or it may apply for loans to bilateral or multilateral lenders, including the IMF Compensatory Financing Facility which, since 1981, has been extended to provide for assistance to members experiencing balance of payments difficulties caused by an excess in the costs of specified cereal imports. This Facility permits countries to borrow an amount equal to the excess of cereal imports above the trend value for a given year, unless this is offset by export earnings above the trend. Interest rates are concessional but the credit has to be repaid in three to five years.

Food aid can play a useful part in food security. Food aid can, of course, also play a part in programmes with objectives other than food security, such as employment generation through rural public works, or programmes of improving nutrition.

There are at least seven criticisms that have been made of food aid, other than emergency famine relief:

First, it reduces the pressure on recipient countries to carry out policy reforms, especially with respect to producer incentives and nutritional objectives.
Second, it tends to depress domestic farm prices, to discourage

domestic agricultural production and to reduce the spread of production-increasing agricultural technology.

Third, it is unreliable, because it depends on donors' surpluses. When needs are greatest, i.e. when prices are high, it tends to dry up.[2] Moreover, since donors make their allocations in terms of money, higher prices buy a smaller amount of grain.

Fourth, if administered through state agencies, it is said to reinforce state hegemony over people and does not reach the poor.

Fifth, it promotes an undesirable shift in consumption patterns away from staples and towards wheat and wheat flour.

Sixth, it disrupts international commercial channels.

Seventh, it leads to unfair burden sharing between donors, if the price of food is overvalued.

The principal objection, that it discourages domestic agriculture by depressing prices can be met by using the counterpart funds from the sale of the food aid at market-clearing prices to make deficiency payments to the farmers who would otherwise be injured, so that supply prices are restored to the level they would be without the food aid. (Even food distributed free, say in schools, frees budgetary revenue if the government would otherwise have paid for it.) In this way the amount by which expenditure on food aid reduces demand for domestic food is channelled back to the farmers and incentives are fully restored. The reason why this obvious solution has not been adopted more frequently is again the budgetary/political constraint. Financially straitened governments normally find other uses for the collected revenue of greater importance and cannot, or do not wish to, collect additional revenue.[3]

Food aid can also be used to finance additional food consumed by construction workers on infrastructure projects for agriculture. Or food aid can be linked with other forms of agricultural assistance to avoid neglect of agriculture. Or additionality of demand can be ensured by distributing the food or its money equivalent to the poorest households who could otherwise not afford it. But the importance of the charge has been greatly reduced, if not entirely eliminated, by the fact that many developing countries have become substantial food importers. (Only in low-income African countries is food aid increasing as a proportion of food imports.) In such a situation the traditional roles of food aid and financial aid are reversed. Food aid, in so far as it replaces commercial purchases, becomes fully convertible foreign exchange, whereas financial aid often remains tied to procurement, commodities or projects. It has, however, been argued that the free foreign exchange made available to governments presents an obstacle to fundamental reforms, such as devaluation of the exchange rate, or investment and reforms in agriculture, which would raise food production. But this is not an argument against food aid, but against all forms of intergovernmental aid. It can be used either to support or to delay reforms.

Food aid can be used either as balance of payments support or as budgetary support. The two extreme cases are, first, where the food aid is wholly additional to commercial purchases and is sold by the government in open markets at market clearing prices, yielding government revenue in the form of counterpart funds of the maximum amount; or, second, where the food aid wholly replaces commercial imports and the foreign exchange saved is used to buy other imports, or more food, or to repay debt.

Historically, there are many instances of food aid that did not harm domestic food production. Forty per cent of Marshall Aid consisted of food aid, yet European food production flourished, excessively. Similarly, South Korea, Israel and India[4] received large amounts of food aid, without apparent long-term harm to their agriculture.

The charge of disruption of commercial sales is greatly reduced by the shrinking and now small role of food aid in total world food trade. If food aid wholly replaces commercial sales by the donor (the government pays the farmers what they otherwise would have earned) no disrupting effects on sales by other countries are suffered. Ensuring additionality, e.g. by linking it with job creation for poor people who spend a large portion of their income on food, also reduces the damage to commercial sales.

Additionality of supply is also important in order to meet the charge that advanced countries that are commercial food importers are faced with higher prices than if, in the absence of food aid, the food had to be sold through commercial channels, lowering prices. The valuation of the food aid has to be done in such a manner as to ensure fair burden sharing between food surplus donor countries and food importing donor countries.

Another charge against food aid is that tastes depend, to some extent, on relative prices and food availabilities (and are not given exogenously, as is often assumed in economic analysis). A prolonged policy of finer grain imports changes tastes away from domestically produced food stuffs and, it is alleged, increases dependency on foreign supplies. The situation has been described as analogous to drug addiction, countries becoming 'hooked' on grain. It should, however, be remembered that these changes in tastes have many causes, connected with development and urbanization, with commercial import policies and with the growing value of time as incomes grow, and that food aid is only one, possibly small, contributory cause.

The volume of food aid has been greatly reduced in the last twenty years.[5] Africa has benefited at the expense of Asia, and within south Asia Bangladesh at the expense of India, and project and emergency aid have replaced bilateral programme aid.

At the same time, so called subsistence crops such as sorghum, millet, yams, cassava and bananas could be traded in local and even national markets, if they were not discriminated against. Low prices of subsidized grain, the import of which is encouraged by overvalued exchange rates, or which is supplied by food aid, discourage the production of these 'poor man's crops' for the market. Although devaluation would encourage the production of export crops, the demand for the subsistence crops would also rise and would constitute an incentive to produce more. The precise amount would depend on the elasticities of substitution in supply and demand. Relatively little research is done on these crops, although there are some exceptions, such as sorghum in Maharashtra and the Sudan, and maize in Zimbabwe. The International Institute for Tropical Agriculture in Ibadan (Nigeria), which is part of the system set up by the Consultative Group for International Agriculture Research, specializes in research on roots and tubers. But more could be done for these crops, especially millet and sorghum.[6] Even where research on food crops has been successful, African countries lack the indigenous research capacity to adopt and adapt the results of this research, so that much expenditure on research has low yields.

To give greater encouragement to research on subsistence crops would have the advantage that they can be grown on marginal land, do not require a sophisticated technology or complex skills, are ecologically benign, and have frequently great nutritional value. They can also be used to supplement the more preferred cereals when these are in short supply, through additions to wheat flour or maize meal. But even if research in this area were to yield good returns there are limits to what can be expected. These crops, particularly roots and tubers, are bulky and expensive to transport. Storing and processing them is costly and often capital-intensive.

The various criticism that have been advanced against food aid have led to the recommendation of better alternatives. Among these is a financial insurance scheme. Countries would then be able to buy food in commercial markets, and not be dependent on the political vagaries of donors. The International Monetary Fund's Compensatory Financing Facility mentioned above was extended in 1981 to apply to cereal imports. The criticism of unreliability of supplies can also be met by multiyear commitments of grain at flexible delivery. These can be bilateral or by groups of donor countries.

10 Price Stabilization

As we have seen, price policy is concerned not only with the *level* of food prices but also with their *structure*, their *predictability*, and their *effectiveness*. In this section we are concerned with their *stability*.

For the poor food producer and food consumer it may be preferable to have stable prices (if they also stabilize incomes or consumption) rather than fluctuating ones, even though the average of the fluctuating prices is higher for the producer or lower for the consumer (as long as price stability does not reduce the incomes of poor food producers in times of bad harvests). The poor buyer of food cannot afford having to pay very high prices at some periods out of his meagre income, even if they are more than compensated by low prices at other periods, for he is not able to carry savings forward. The small farmer who lives on the margin of subsistence cannot afford to accept risks that might drive him below that level, even if they are accompanied by prospects of substantial gains. This is one of the factors preventing subsistence farmers from switching to cash crops. It is the presence of such risks, and farmers' aversion from them, that prevent optimum resource allocation in competitive markets.

One way of overcoming the obstacle set by the fear of falling below minimum earnings is to provide crop insurance so that the possibility of disaster is eliminated and risks become acceptable. But such crop insurance has been largely a failure in developing countries (and is very heavily subsidized even in developed countries, though less than in developing countries). The reasons for the failure are obvious: moral hazard (farmers neglect to take the actions required to reduce the risks), adverse selection (only the high-risks farmers insure their crops), high costs of administration (typically 6 per cent of the value of coverage compared with about 1 per cent for life insurance, raising premiums to 20 per cent of coverage), inefficient management and inability to reach the small farmer. This last defect is both inequitable and inefficient, since it is the small and poor farmer who is most risk-averse and who should be induced to adopt higher yielding but riskier technologies. In order to contain moral hazards, inspection and monitoring costs rise. If these are kept down, indemnity payments rise. Where farmers rely on credit – and again these are normally the larger and richer farmers – crop insurance amounts to insuring the banks, as for example in Brazil. The only exceptions to this poor record are very specific insurance schemes, such as those against hail, flood and hurricane damage, where the moral hazards are minimized.[1]

It is better to eliminate the causes of the risk in income by measures such as reliable irrigation or reliable fertilizer deliveries. But not all risks can be eliminated by government action. Crop insurance is mainly against shortfalls in yields, but risks also arise from bad health, shrinking markets and failures of supply of inputs. The general lesson of the experience with crop insurance is that alternative ways of reducing risks for small and poor farmers should be explored before adopting crop insurance schemes.

Another way is to develop futures markets which would reduce at least the risk of variation in price, though not in output. A guaranteed, stable supply price is also a greater incentive to investment and to higher production than a fluctuating price, whose average is higher. It is equivalent to a reduction in the costs of production or the costs of borrowing, and enables the farmer to increase his investments beyond what they would otherwise have been. One exception to this would be the case where poor farmers' principal concern is to be sure not to fall below a certain minimum income, necessary for survival. They will work very hard, and long hours, to secure this aim. If an insurance scheme then reduces the variations in their income they may react by producing less.

A particular aspect of instability in food prices and food supplies is seasonal variations. Immediately after the harvest food is plentiful but may have to be sold at low prices, sometimes in distress sales to repay debt. Between harvests food may become scarce and malnutrition may rise. Farmers may have to sell the little plots they own to sustain consumption. Price stabilization to overcome the seasonal aspects of poverty and malnutrition can be an important objective. It is true that such stabilization will reduce the incentive of farmers to store food and carry it from times of plenty to times of scarcity. And, since their income after the harvest will be increased compared with a situation of fluctuating prices, they would have to save some of it in order to buy, though at lower prices than would otherwise prevail, in times of scarcity.

Famines can be caused by sudden large increases in the prices the poor have to pay for their food. Even where such increases cause lower food prices in the long run, the poor have to be cushioned in the transition period, a problem discussed in a separate chapter.

Price stabilization is also beneficial to non-poor producers and consumers, even if the level is lower than the average of fluctuating prices, for producers, or higher for consumers, if the agents concerned are risk averse.

Stabilization at costs not exceeding the benefits against short-term fluctuations in demand is often regarded as beneficial,[2] for such short-term fluctuations are at best useless, at worst harmful to the improved allocation of resources. Uncertainty discourages efficient planning of production in the industry concerned, and efficient planning of use by the buyers of the commodity. On the other hand, *stabilization* of prices in the face of *fluctuations in supply* (as is often the case with agricultural crops subject to the vicissitudes of the weather or for other raw materials as a result of civil disturbance or war, or of technological changes in inputs), destabilizes incomes and worsens the producers' plight, unless demand and supply elasticities are sufficiently low, in which case price stabilization also stabilizes incomes for products not traded internationally.[3] This may well be the case for many agricultural commodities. But for higher elasticities price stabilization would destroy the compensating effect that price movements have on output movements when supply fluctuates.

There are also other reasons why price stabilization may not stabilize incomes. If farmers produce and sell a range of crops, price and income variations in one crop may compensate for opposite variations in other crops. In that case stabilization would reduce stability of total income. Or producers may be insulated from price fluctuations through marketing boards, which would then benefit from the stabilization rather than the farmers. Or the variation in price of the goods bought by the farmer may be in the same direction as the variation of the price and money incomes of the crops he sells. In that case, real income may be destabilized by price and money income stabilization.[4]

It is important to ask not only whether price stabilization leads to real income stabilization, but also whether incomes are raised or reduced. For shifts in the demand curve, price stabilizes producers' incomes, but also reduces them. For shifts in the supply curve, price stabilization can destabilize producers' incomes, but also increases them. The question is whether the benefits of stabilization outweigh the reductions in income.

Finally, there are the effects of price and income stabilization on investment and on improved economic planning and policy-making. These are difficult to measure precisely, but they are likely to be more important than the static gains, which can be more easily captured in formal models.

There is a variety of ways in which the prices of primary products, and with them the incomes of their producers, can be *stabilized*, in the face of demand fluctuations, which score different marks with respect to equity and efficiency.[5] One of these is *buffer stocks*. Given adequate foresight, independent administration, resistance to political pressures, and low storage costs, the operators of buffer stocks could be, like Joseph in Egypt (in words of Tom Lehrer's dope pedlar), 'doing well by doing good'. Unfortunately, none of these conditions are given. None of us is granted perfect foresight by the Almighty, as Joseph was in his dreams. Buffer stocks, as we have seen, are also expensive. It will be remembered that in the Sahel, for example, one ton of buffer stock grain might cost as much as $500, compared with $200 per ton for the world c.i.f. price.[6]

For any proposed buffer stock scheme, we should also have to make allowance for its effect on private stocks held by agricultural commodity dealers. These stocks and trading in futures markets would be reduced if successfully operated official stocks went to replace them. To that extent, the benefits of a new buffer stock are reduced, unless the official operators have better foresight or operate at lower costs than the private ones.

If the official stocks are badly operated, so that they sell in times of plenty and buy in times of scarcity, well-managed private stocks would, of course, have additional scope for operation. Since badly operated stocks lose money, while stockpilers with correct guesses make profits, one would expect the badly managed stocks to be driven out of business. But this need not be so either if they are subsidized out of public funds or if the members of the losing stocks are a continuously replenished group of outsiders, attracted by the profits of the insiders, while the insiders stay the same. It would therefore be possible for a group of insiders to continue to make money from speculating on the mistakes of the outsiders, without paying any attention to the real forces. In this way buffer stocks could continue to destabilize prices indefinitely and be harmful to both producers and consumers.

The success of public buffer stocks will depend upon adequate margins between the floor prices paid to farmers and the ceiling prices paid by consumers. Political pressures will be exerted on both: there will be farmers' pressures to drive buying prices up and urban consumers' pressures to keep selling prices low. If these are resisted, if there is reasonable foresight and if there are adequate administrative and financial resources, such stocks can be socially useful.

The main problems with buffer stocks arise from the difficulty of correct price predictions, the political pushes and pulls of buyers and sellers – producers want stabilization at a high level, consumers at a low level – their failure to stabilize earnings by stabilizing prices in some conditions, and the high costs of operating them.

Other measures to reduce price fluctuations and risks are multilateral long-term contracts, the provision of information about future prices, the use of imports and exports to stabilize domestic prices, futures markets (in which the price for the whole crop is fixed now, irrespective of what it will be when the sale occurs) and option markets (in which the seller buys the option to sell at a certain price in the future, but does not have to do so).

National, regional, and international stocks and buffer stocks, foreign exchange reserves, access to credit on appropriate terms, and other measures such as options and futures markets, can be both substitutes for one another or complementary to one another as paths to stabilization. Each serves similar or different purposes. Foreign exchange reserves and access to credit are best suited for local harvest failures, though some local stocks are also needed. Foreign exchange reserves have the great advantage over national buffer stocks that they are much cheaper to hold and more flexible in their use. But they depend on the ability of developing countries to export and earn reserves, which, in turn, depends upon trade liberalization by the industrialized countries, and greater stability in export earnings. Foreign exchange reserves are less suited to meet worldwide harvest failures. The country would then have to buy at greatly inflated prices and might be faced with non-availability. Ability to draw on international (or regional) buffer stocks could mitigate or eliminate this problem though it might be cheaper to maintain excess productive capacity, e.g. in the form of irrigation water. Buffer stocks may also be indicated when the commodity is not traded in international markets, or not on an adequate scale, such as edible beans in Brazil.

The income stabilization, which is one of the main purposes of price stabilization through buffer stocks, can be achieved either by supplementing it with a crop insurance based on volume (although we have seen that this suffers in practice from various defects) or, without price stabilization, with an insurance based on total value. Guaranteeing incomes to individual farmers would not work because it would remove the incentive to produce anything, but relating guaranteed prices to total output would be a step towards income guarantees, if each farmer contributes an infinitesimally small share of that output. The development of futures markets would remove the risks of fluctuations in the price component of incomes and would leave the farmer free to choose whether he wants to bear the risks or not, but would leave output fluctuations unaffected. Bulk purchase agreement and long-term contracts would shift the risks from the producers to the buyers.

More important than these microeconomic effects of price stabilization[7] is likely to be the macroeconomic impact on worldwide inflation and the effects of the consequential anti-inflationary measures. The analysis is based on a fundamental distinction in the behaviour of primary and manufactured products. Food, raw materials and energy (land-based supplies) are inputs into the production of manufactured products. If the production of the latter races ahead, the prices of primary products rise and brake the manufacturing advance. As manufacturing decelerates or is reversed, prices of primary products fall and eventually stimulate growth again. Since 1971 the prices of primary products have become much more sensitive to variations in industrial production than they were before.[8]

In view of institutional forces such as real wage resistance to declines when prices rise and oligopolistic market structures, a rise in commodity prices will tend to raise the prices of manufactured products, but a fall will not reduce them. It will lead to a drop in production and employment instead. This rise can become cumulative as a result of speculation and a consequential wage–price spiral. Such asymmetrical behaviour will give an inflationary impulse to commodity price fluctuations.

Various price stabilization schemes have been introduced, among them the Common Agricultural Policy of the European Economic Community (EEC) which, as Lord Kaldor has argued, has two advantages. First, the guaranteed price amounts to a cost reduction for producers and encourages investment in agriculture. Second, by raising agricultural incomes, it raises the demand for industrial goods and speeds up the growth of both sectors.

But being only a geographically partial scheme, the benefits to the Europeans are bought at the expense of the rest of the world. The EEC policy of the variable levy, which stabilizes EEC prices, destabilizes them for the rest of the world. As was argued above, the variable levy reduces elasticities of both demand and supply. When there is a world surplus and world prices are low, the levy rises, thus reducing world demand and adding to world supply, aggravating the surplus and lowering further world prices. When there is a global shortage, the levy falls, adding to demand and subtracting from supply, thereby aggravating the shortage and adding to world price increases.

In the recent past, price increases caused by commodity price rises have been countered by monetary and fiscal restrictions in the industrialized importing countries. It was thought that 'accommodating' monetary policies would be inflationary and the restrictive policies then reduced the demand for imported raw materials and their prices. Reductions in demand and income in the industrialized countries take the form of reductions in output and employment. But since the unemployed receive unemployment pay or other forms of assistance, the hardships are not too great. And they are mitigated for the developed countries by improvements in their terms of trade as a result of the fall in the prices of imported raw materials. For the developing countries, on the other hand, the reduction in demand and the drop in their export earnings can be quite disastrous. People living precariously on the margin of subsistence are driven below it. This disagreeable dilemma between, on the one hand, import cost-induced inflationary price movements (aggravated by speculation and money wage increases), and, on the other, underutilized capacity, lost output, unemployment and depressed earnings for developing countries from commodity exports, can be eliminated or reduced by a commodity price stabilization scheme.

Such a scheme would have the principal function of 'aligning the growth of world industrial production to that permitted by the growth of availabilities of primary products . . . not through price variations but through variations in the rate of investment in stocks . . .'[9]

The buffer stock, which would contain food but not be confined to food, would buy commodities whenever the increase in the supply of commodities exceeded requirements of industrial production, thereby offsetting the otherwise deflationary impact, and it would sell when industrial production exceeded the supply of commodities, thereby dampening the otherwise inflationary impact. Purchases should be financed not out of taxation but should be regarded as capital expenditures, and sales should not be accompanied by reduced taxation. A good way of financing purchases would be the issue of new Special Drawing Rights (SDRs). In this way the buffer stock would contribute to providing the world with a stable unit of international money.

It must, of course, be noted that the achievement of these macroeconomic objectives has no direct bearing on the main concerns of this book: raising producers' incentives and safeguarding consumers against price increases. Balance of payments objectives can be achieved, inflation and unemployment reduced, and steady growth be attained without making either poor food producers or poor food consumers better off. The only tenuous link is that the achievement of the macro-objectives should make it easier to pursue policies that benefit the vulnerable groups.

Newbery and Stiglitz have raised questions about the macro-economic argument. They argue, first, that exchange rates could vary so as to adjust the prices of manufactured goods in international markets and compensate for their rigidity in home markets; secondly, that exchange reserves could provide a buffer for fluctuations in trade, and third that countercyclical government policies could overcome or reduce the cycle of inflation and unemployment. But the macroeconomic case for price stabilization remains to be more fully explored.

11 Export Crops *v.* Food for Domestic Consumption

There has been a good deal of discussion of the respective merits of export crops and food for domestic consumption. Some have argued that if the comparative advantage points to export crops, it is they that should be promoted and the foreign exchange earned be used for inputs into agriculture or industry, or even for food imports. On the other hand, there have been those who have argued that export crops impoverish the poor, deprive the people of food, and lower nutritional standards. It has also been argued that they are ecologically destructive.

A parallel debate has gone on over the respective merits of marketing more food, whether for the domestic market or for exports, against growing more food for consumption within the farm household that grows it. Some of the arguments that apply to the debate on exports versus food for domestic consumption also apply to the debate on more marketing versus more production for own consumption.

Clearly much depends on the institutional arrangements. If export crops are grown on large plantations, perhaps owned by foreigners, which generate little employment, while food is grown by small farmers, the impact on income distribution will be different according to which type of crop is promoted. If the foreign exchange earned by export crops accrues to the government and it spends it on arms or office buildings, while the receipts from food would have gone to the poor peasants, again the distributional impact will be different.

The interesting fact, at least in Africa, is that there is little evidence that there is a necessary conflict between the two. Where land and labour are not scarce, the movements in export crops go in the same direction as the movements in the production of food for domestic consumption. There is also little evidence that nutritional standards have suffered as a result of the growth of export crops, though sugar in Kenya may be an exception. There are several reasons for this. First, certain services can be in joint supply and help both export crops and food. Among these are extension, marketing, and supplies of inputs.

Similarly, equipment and fertilizer can be used to raise the production of both. It can also be the case that similar complementarities exist on the side of demand: the demand for food by the export crops growing farmers creates a stable market for local food supplies and encourages their increase. For some export crops, such as coffee, the opportunity costs are very low. In other crops, such as cotton, a part of the export crop can be used as food, in the form of edible oil, or as feed in the form of cottonseed cake.

In spite of the positive relationship between export crops and food in some conditions, clearly conflicts can arise. Where there has been in the past unwarranted discrimination against food crops, there is a case for removing that discrimination. Colonial governments have tended to favour export crops by improving infrastructure, such as transport, marketing and distribution services, compared with facilities for domestic food consumption. Research activities and extension services also have been concentrated on export crops, particularly if linked to large firms with market power and capacity to process the crops and make profits by raising prices in the face of an inelastic demand. On the other hand, export crops have often borne higher taxes than food crops and are more easily controlled by governments, since often they grow in specialized regions, they have to pass through ports, the buyers are more concentrated, etc. But economists have a bias in favour of trade, whether international or intranational, and against locally consumed food. Such historical biases, where they exist, should be removed.

Some scholars have gone further and have criticized colonial governments for destroying the integrated farming–herding systems which, in precolonial times, protected the ecology while allowing substantial food production. In Africa, in precolonial days, farmers opened their fields in the dry season to pastoralists who brought their herds to graze on the harvested millet and sorghum stalks. The animals were fed through the grazing, they manured and fertilized the soil with their dung, and the cattle hooves broke up the earth around the plant stalks, allowing oxygenation of the soil. The herdsmen traded their milk for the farmers' grain, and people, animals and land were simultaneously maintained. The introduction of monocrop cash production (peanuts, cotton) by the colonial governments destroyed this system. In addition, well digging concentrated animals round watering sites and their trampling turned them into small deserts. Post-colonial production further encouraged this erosion of land, and in the view of some contributed to the present famines.[1] It is, however, not clear how this system could have been maintained in the face of a rapidly growing population.

In some cases export crops face an inelastic world demand. Then production should be curtailed and total proceeds raised, unless current price rises reduce future demand, appropriately discounted, by more than present gains. Such reasoning underlies the attempts to design commodity agreements for tea, cocoa, coffee, sugar, spices, etc. Much current research is devoted to raising further the productivity and production of such export crops, not always to the benefit of the growers. The difficulty here is that, in the absence of effective commodity agreements, the national efforts of small countries are not coordinated with those of the growers and exporters in other competing countries.

A shift to export crops sometimes reduces the role of women who traditionally produce, prepare and distribute food. In spite of the higher family income earned, this can lead to a reduction in nutritional standards. But few generalizations are possible. Sometimes women increase their labour in food crop production to compensate for the reduced labour of men, sometimes producing surplus output for sale, thus raising both their income and their independence. Sara Berry reports of a case in Cameroon[2] where men took up cocoa, coffee and bananas, abandoning the food farms they had previously cultivated. Women took up the slack. But the women grew the food crops in an entirely different way from the men, using a system of cultivation which involved small, daily outputs of labour throughout the growing season, in contrast to the men who had cultivated a combination of crops, requiring occasional short peaks of concentrated labour. The result was an increase in total output, of export crops and of food crops, where the additional food was grown by fewer people (only women) who used more labour-intensive techniques than the average labour intensity of men and women combined in the past.

The impact of changes to export crops on (a) expenditure patterns as a result of cash accruing to different members of the family, and in large, discontinuous lumps rather than as a steady flow, (b) distribution of food to different members of the family and (c) the allocation of time and effort by women to different types of work are important areas of study. On the other hand, a shift to export crops may raise employment opportunities and therefore offer more people access to food, as in the case of a shift from rice to jute in Bangladesh.

A good deal depends on the distribution of land and the mode of agricultural production. If export crops benefit large plantations or commercial farms, whereas food is mainly produced by small farmers or their wives, a switch to export crops can aggravate inequality in access to resources, income earning power and land ownership.

(Examples are sugar in Jamaica and cotton in El Salvador, but even in Africa the trade bias has favoured the large export-orientated farm and firm.) Export crops are often grown more efficiently in large farms, and the change-over can lead to an impoverishment of small farmers. In African countries, additional export earnings tend to increase the income of small farmers, and the extra foreign exchange contributes to the ability to break bottlenecks in transport and agricultural inputs, which are important for food production. This was certainly the case in Tanzania during our ILO mission in 1981.[3]

How much should be devoted to export crops is also determined by trends in the costs of international transport, which in turn is affected by oil prices. The lower these costs, the stronger the case for international trade. Expectations of higher future transport costs, other things remaining equal, would justify a move towards reduced dependence on foreign trade.

Foreign exchange is often one of the scarcest resources, while its increase can make fuller use of many domestic resources possible and contribute to greater food production, as well as to higher imports of food itself. It is sometimes possible to increase the production of both export crops and food for domestic consumption, particularly if improved technologies are introduced. In some cases the opportunity costs of increasing exports are very small, and the choice does not arise. Land may be plentiful, and little labour and other inputs be required. Where domestic food production does decline, it is possible to encourage home gardens simultaneously with the expansion of export cash crops to ensure continuing adequate nutrition. A Kenyan Ministry of Health study of 1979 showed little evidence that four export crops (coffee, tea, cotton, pyrethrum and sugar cane) had been detrimental to nutritional status. The only possible exception was sugar cane.[4]

However, some qualifications are needed to the notion that resource allocation should be guided by comparative advantage, so that a comparative advantage in an export crop can buy more food from abroad than would be produced at home. The comparative advantage is not God-given but itself determined by the direction of research, and research has been heavily biased in favour of export crops and the staple grains, to the neglect of 'inferior' food such as millet, sorghum and cassava. With the growing importance of human capital, the direction of comparative advantage can be quite strongly influenced by research, extension services, education, and other forms of investment in human beings. It is no longer only or even mainly 'endowments' that determine specialization in international trade, but conscious policy decisions.

Recently, there have been some successes in research on these 'inferior' food crops. In Zimbabwe hybrid varieties of maize, in the Sudan high-yielding, drought-resistant strains of sorghum, and in Nigeria a disease-resistant variety of cassava with three times the yield of native strains, have been developed. But a complete elimination of the bias in favour of export crops, combined with provision of credit and delivery systems would change the comparative advantage and convey benefits to poor people. There is also some evidence that poor people produce the things they themselves consume, and consume the things they produce. There are several reasons for this. First, when households switch from semi-subsistence cropping for their own needs to monocropping for export, their income may rise but their nutritional status drop. Second, monocropping for export may raise the risk of crop failure, even though the average returns are higher. Third, export crops often take a long time to mature, and the outcome may turn out to be less profitable than expected. Fourth, there is a distributional consideration in favour of growing food. There are two dangers in simply raising the productivity of the poor by switching to export crops.

First, there may not be adequate demand for the things they produce, or export taxes may be levied, or marketing margins may be high; and, second, the price of food, on which they will want to spend a large part of their extra income, may rise sharply, particularly if the change involves shortages of food in local markets. These two dangers are more likely to be avoided if the poor can meet their own needs in somewhat more self-sufficient units than would be indicated by a strict application of the theory of exchange and comparative advantage. This applies to families and households, to villages, to nations and to groups of poor nations. There are distributional advantages in this mutual meeting of basic needs which have to be set against the conventional claims of the aggregate gains from trade, which may be greater but less well distributed, more uncertain, or longer delayed.

In some countries such as Zambia, Mali and Tanzania, a dilemma arises. The above arguments for encouraging smallholder production of food for local consumption are strong. At the same time, foreign exchange scarcities constitute a bottleneck to expansion because they reduce the availability of consumer goods, fuel, transport equipment and fertilizer. If productivity in food production is to be raised, growth of inputs is necessary, and this frequently depends on importing these inputs. Productivity growth depends crucially on moving towards machinery, fertilizer and pesticides which often have to be imported and cost foreign exchange. Imports have been scaled down to the minimum, so that, without extra aid, an increase in exports is the only solution. Local food production and consumption cannot be raised without raising exports, but exports can be raised only by curtailing food for local consumption. Non-project, untied foreign aid combined with the right policies can transform this vicious circle into a virtuous one.

Generalizations, such as export crops are grown on large farms, food crops on small; export crops are grown as monocrops, food crops in diversified farming enterprises; or export crops are more or less labour-intensive, or use more or less female labour, are quite impossible to make. Some progress has been made by combining output, sales and methods of production in different ways. The best guideline is to avoid dogmatism on this issue but to promote policies that raise and stabilize the incomes of the poor people, whether through exports or food for domestic consumption or both, and to make sure that they have access to the food.

To sum up the controversy: the passionate opponents of export crops, the value productivity of which is often higher than that of food production, have, on the whole, not provided good reasons for their attack, but there is a kernel of truth in their criticism of the advocates of comparative advantage as a guide to foreign trade. This can be summed up under the following headings.

1. Comparative advantage can change, particularly as a result of changes in the direction of research and human capital formation.
2. The institutional arrangements as to who benefits from foreign sales (government through export taxes, parastatals, foreign firms, plantations, large commercial farmers) and from domestic production of food (small farmers) make an important difference.
3. Local production and local markets for food can be harmed or destroyed by foreign trade. In spite of higher earnings to the country, local food prices may rise or certain foodstuffs may cease to be available. But local self-sufficiency in food, like national self-sufficiency, may also harm the poor.
4. Higher incomes to the growers do not always mean that nutritional standards of all members of their families are improved.
5. Export crops sometimes carry higher risks in production, the costs of foreign transport, and foreign demand.

6. The distribution of benefits (and power) between men and women, and between government and private agents, may be different.
7. Foreign trade contributes to a change in tastes which can make the country both more vulnerable and reduce nutritional standards.
8. Monocropping for export can be ecologically harmful.
9. In many situations experience has shown that food and export production are not alternatives but complementary.

12 Income Distribution and Poverty

A rise in food prices reduces the real income of poor food buyers by a larger proportion than that of better off consumers, though the absolute reduction in real income is larger for the higher income consumers, because they spend more on food. This reduction in turn will reduce employment and income of the poor in those sectors on which the expenditure of the better off is reduced. The poor therefore suffer both directly, as a result of the increase in food prices, and indirectly, as a result of reduced employment in the production of e.g. livestock and vegetables, simple household goods, kitchen ware, bicycles, etc., because the expenditure of the better off has declined.

John Mellor has calculated that in India a food-grain price rise of 10 per cent reduces the total real expenditure of the lowest two deciles by somewhat more than 10 per cent.[1] (This does not take into account the fact that the beneficiaries of the higher prices may spend their higher incomes on goods and services that raise the incomes of the poor, because this will take time.) In contrast, higher prices for crops tend to benefit disproportionately larger farmers who account for a large share of marketed output.

To vary a remark that Scott Fitzgerald made to Hemingway, 'the poor are different from the rest'. They consume different types of food, they live in different places, they have different livelihoods and jobs, they have different-sized families. Their ratios of consumption to income, of food to consumption, and of cereals to food consumption, are higher than those of better off people. Consumption is different according to geographical region, age, and sex, and, above all, season. The incomes and the access to food depend on different factors for different social and economic groups, and for different time periods.

It is useful to distinguish between four vulnerable groups: (1) subsistence farmers, (2) deficit farmers and landless labourers, (3) small surplus farmers and (4) the urban poor. The access to food of each of these groups depends on several factors. The four groups are distinguished by their functions, not in terms of their composition by individuals or households. It is important to remember this because, as

60

a result of price policies, the composition of the groups is likely to change.

1. While pure subsistence farmers are rare, since most sell something in the market, if only their labour, and have some cash commitments, if only to repay debt, the expression is used here for those who rely principally on subsistence agriculture. The World Bank estimates their number to be approximately one billion. That part of their crops which is consumed by the household is, by definition, insulated from market forces, although it may vary in size according to the prices that can be received from market sales. That income depends solely on the farmers' productivity, which is usually low, and price policy has no effect on it.[2]

2. The access to food of landless labourers and deficit farmers,[3] who have to hire themselves out for some of the time and earn wages, depends on employment, wages and the price of food (or, if paid in kind, on the amount of food they receive). Similarly determined are the incomes of the rural informal sector's self-employed: (small-traders, potters, porters, blacksmiths, carpenters, leather-workers and van drivers.

3. The incomes of the small surplus farmers and of nomadic herdsmen who have to buy food depend on their productivity, the prices of the crops and food they sell, and the prices of the food they buy.

4. The incomes of the urban poor depend on employment opportunities, wages and the price of food. But the urban poor in some countries (e.g. Africa) have rural links and therefore can either return to their villages and join the ranks of the rural poor or get food supplements from the countryside if their access to food in the town is reduced. This was certainly true in colonial times but is much less true today, since rural areas have suffered from declining output as a result of ecological damage, adverse policies, heavy migration, excessive commercialization, etc. The economic benefits of rural links have greatly declined. The 'homelands' in South Africa are an extreme example of areas where the flow of 'benefits' has been reversed.

The impact of a rise in the price of food is different for these different groups, and the responses are different in the short run, medium run and long run. It should be stressed again that the composition of these groups may change as a result of a rise in the price of food so that, for example, subsistence-orientated farmers or deficit farmers may become surplus farmers, or, alternatively, landless labourers may become urban poor, etc. This may sound somewhat inconclusive but no dogmatic generalizations can be put in its place. What is needed to assess the distributional impact is a careful analysis of each separate vulnerable group and how it is affected by a rise in the price of food. In different countries the relative importance of these groups will differ. In south Asia, for example, there are many rural landless labourers who have to buy food. In sub-Saharan Africa, on the other hand, the poor, at least in the past, have largely been sellers of food who would benefit from higher prices. In Latin America, the urban poor who have to buy food are numerous. In many low-income countries the poor are largely subsistence-orientated farmers, so that changes in food prices will leave the subsistence part of their production unaffected.

Subsistence-orientated farmers tend to be low-productivity farmers (Fiji and Papua and New Guinea appear to be the exceptions). A rise in the price of the food crops which they produce has two opposite effects. On the one hand, there is a tendency for the household to reduce demand for its crops and sell more to the (official or unofficial) market. On the other hand, because its real income has risen, there is an opposite tendency to demand and withhold more of the crops for consumption by members of the household. The net effect depends on which of these two tendencies is stronger. In addition to these changes in demand there are also opposite effects on supply: more hours are worked because the reward has risen, but fewer hours because with higher income leisure becomes more important. And there may be a switch between crops, with different labour requirements, if their relative prices change.[4]

The cessation of low grain prices, resulting from subsidies, imports at overvalued exchange rates or food aid, together with some encouragement through increase in research expenditure, may convert a part of their subsistence crops into nutritious, marketable crops. This would have the advantage of making use of marginal, low-quality land with lower investment requirements than grain or rice cultivation, and meeting the nutritional requirement of the rural population, and in some countries of Africa, also part of the urban population with rural links. On the other hand, farmers may switch to producing more grain and less of the subsistence crops.

Policies will have a different impact in Africa where there are few landless labourers, from south Asia, where there are many; and in countries of south Asia, where there are many subsistence-orientated farmers who sell only a small share of their crops, from those of South

America where the marketed surplus plays a greater role.

A rise in food prices will leave that part of the food crop that is retained by the household unaffected. *The landless and deficit farmers* will suffer a reduction of their real income in the short run, but in the longer run employment may increase, both on the farm as a result of grain production having become more profitable, and off the farm, as a result of surplus farmers having more income to spend. Wages may rise as a result of the increased demand for labour. If labourers are paid cash, the short term loss may be partly or wholly offset by this rise in money wages. If they are paid in kind, their real income is unaffected. But it is not uncommon that the method of payment changes from kind to cash when prices rise, as commercialisation increases, and that the increase in money wages does not fully compensate for the rise in food prices. If the higher price of grain converts deficit farmers into surplus farmers, because they now work more on their own land, they also gain. If the farmer is sometimes in surplus and at other times in deficit, and if he has to borrow when in deficit, his position will depend in addition on the terms of the loan.

A steep rise on food grain prices may force *deficit farmers* to sell some assets, including land. If they have borrowed in the past and pledged their land as collateral, and if default occurs, they lose their land. As we have seen, it is then possible that the loss of productivity per acre of the small farm is not offset by a faster rate of adoption of new technologies by the large farm, so that the price rise leads not only to impoverishment but also to lower agricultural production. Something like this may have occurred in Bangladesh. It is certainly not common that agricultural output declines when prices of crops are raised. The illustration is given only to show that such possibilities cannot be excluded and that much depends on the distribution of land holdings and on tenurial arrangements.

It is often said that this is a situation that may be characteristic of south Asia, where land is scarce and labour abundant, and where there are *landless labourers*, but does not apply to Africa. Yet in Africa too, similar vicious spirals can arise. Farmers pledge not their land but their crops, and herdsmen their cattle, when food prices rise and they have to buy food, and are then further impoverished and forced to hire out their labour to repay the debt. The problem is aggravated if the poor are forced into distress sales of their crops immediately after the harvest, when prices are lowest, to repay debts and maintain good relations with their patrons. Prices can be further depressed at this time by monopsonistic power on the part of the food buyers. Even in

Africa there are many rural poor who have to buy food, such as
herdsmen and workers in the rural informal sector. And with the high
rates of population growth there is a growing class of landless
labourers, for instance in Kenya.

The *small surplus farmer* is better off because he gets more for his
crops, even though he may have to pay more for the food he buys. The
urban poor suffer a short-term decline in their real income. The long-
term effects of higher food prices on the urban poor are complex. If
higher food prices improve the lot of farmers and reduce rural–urban
migration, real incomes in the urban informal sector may increase
compared with what they would otherwise have been. Money and real
wages in the organized sector may also rise, particularly if wages are
linked to the cost of living. But if the prices of industrial products rise
with higher wage costs, the agricultural terms of trade have to that
extent not improved.

Some observers argue that when grain or rice prices rise the poor
substitute inferior types of food for the grain or rice and do not suffer
nutritionally. Since the supply of these substitutes is also inelastic in
the short run, the prices of the substitutes will tend to rise. In
Bangladesh, the prices of inferior substitutes rose faster than rice
prices when food-grain prices in general were rising.

It is also often argued that the short-term increase in agricultural
prices would raise supply in the long term and lead to lower food prices
than would otherwise have prevailed. This is not possible as long as the
movement is on the same supply curve, for the increase in supply was
induced by the higher price, and if the price fell, there would be a
return to the lower level of supply. But the short-term increase in price
could lead to improved incentives and ability to invest, innovate, and
adopt technical change. The short-term increase would then lead to a
downward shift of the whole supply curve. This is, of course, perfectly
possible and appears to have happened. The goal is to get food output
rising faster than food prices are falling. Two qualifications are
necessary here. Much depends on the initial system of land distribution
and access to agricultural inputs, innovation, information, institutions,
infrastructure. (Together with incentives they constitute the six 'Ins' as
necessary conditions for agricultural growth.) In a society in which
power and wealth are very unequally distributed, the rise in the price
of food may make the rich richer by leading to transfers of land,
without leading to a lowering of the supply curve. This may have been
the case in Bangladesh. A study of Thailand has shown that an increase
in rice prices has hurt the rural (as well as the urban) poor, particularly
those in the informal sector.[5]

The second qualification relates to the period of transition. Even if all the desirable long-term results were to come about, this would be of little comfort to those poor who would starve to death in the short run. We return to this in the next chapter on the transition.

The impact of food price increases on the poor is, as we have seen, highly complex, but may be summarized in the following way. A rise in the price of food raises the real incomes of food producers and lowers, in the short run, the real incomes of food consumers, since in the short run supply does not rise. There are some very poor people in both groups. In principle, consumer and producer prices can be delinked by a system of taxes and subsidies, but in practice there are severe constraints. In the medium and long run the detrimental impact on poor food consumers can be mitigated or offset by changing technology and a consequential increased supply of food, increased employment, both on and off farms, e.g. from multiple cropping, higher wages, and perhaps reduced rural–urban migration with consequential higher urban wage and income levels.

Many developing countries have dealt with the basic dilemma whether to raise prices to encourage production (and benefit poor food producers) or to keep them low to safeguard poor food consumers by introducing a dual market. The market is divided into two sectors, one in which relatively high prices clear supply and demand, the other in which the government steps in, procures a part of the grain supply either as a tax or at low prices, and sells it at concessional prices to the poor. The effects of such a dual system are that the free market price rises because of the additional demand from poor consumers, which is only partly offset by the reduction in demand by the richer groups because of the price rise. The difficulty is to reach *all* the poor and *only* the poor. If some poor food consumers are excluded from access to the low-priced food, they will be doubly worse off: first because of exclusion and second because of the price rise. When this is the case, measures that raise the controlled price of food will benefit the poor excluded from these markets, because they will tend to lower the non-controlled price. If some non-poor have access, stocks may run out, and either extra imports will be needed or not all poor customers will be served. In such a system, there is a transfer of resources from farmers to consumers of food. Such a transfer would be undesirable if farmers were poor and the consumers middle-income urban groups. Once such a system is adopted, the problem is to run the low-price market for the poor at the lowest social costs. The problems are similar to those discussed in the next section, concerned with the transition, only here they may represent a permanent feature.

13 Protecting the Poor in the Transition[1]

It has been argued that the ultimate resolution of the fundamental dilemma with which we started must be found in productivity growth, and that higher producer prices have an important role to play, in conjunction with the right technological package. Higher yields, higher profits, and higher incomes to farmers and farm-workers can then be combined with greater output and lower prices of food. (For any one country the benefits of productivity growth can be passed on to foreign consumers in the form of lower prices. But the proposition remains true for the world as a whole.) This process takes time, and until its benefits are evident, the poor buyers of food may starve.

Keynes said we are all dead in the long run. This does not mean that we are all alive in the short run. Very poor food buyers may starve to death before the blessings of the long run materialize. In Bangladesh, high rice prices have led to higher child mortality.[2] Policy-makers embarking on a course of raising food prices must therefore pay special attention to protecting the poor in the transition.

The high costs – in terms of political constraints – and economic and financial resources of the transition from bad to good policies is one of the biggest obstacles to reform. Any measures that can ease the transition are an important contribution to better policies. The international implications are discussed in a later section. This section is concerned with the domestic policies a country should adopt to make the transition easier.

It has already been argued that two remedies often taken are not normally very effective. Additional grain imports will tend to reduce the incentive effect of higher grain prices, unless measures are adopted to counteract the damage. Attempting to keep the secondary grain and root crop prices low to protect the poor is not effective if their supply is not elastic. Their supply would be even reduced if acreage were switched to the higher priced grains.

One method of protection is to increase the price gradually, in small steps. If the rise in agricultural output is thereby delayed, this might be regarded as a price worth paying for the nutrition objective. Some would argue that the positive incentive effect is probably as strong for small and gradual as for sudden, and large increases, but poor food consumers are somewhat cushioned. Very large and sudden price increases might even have perverse effects on supply, because producers will take out some of the gains in greater leisure. In any case, the large rents generated for the intramarginal producers do not fulfil any economic function. On the other hand, there is some evidence that the incentive effects are better achieved by sudden and substantial price increases, which alert the farmer to the new profit opportunities. If this shock effect, requiring a rise above a critical threshold were to be important, other ways would have to be found to cushion poor food consumers. It is likely that the assurance that prices will increase steadily is more important than a sudden and large increase. Gradual-ness, combined with confidence, is a better incentive than large changes that are expected may be reversed. It is therefore important not only that the new policy be clearly announced, but also that a firm timetable is set for the period of transition, so that farmers know by how much prices of their crops will increase each year.

A second way is to make the subsidies more selective by concentrat-ing them either on vulnerable groups or on basic food staples consumed by the poor, like cassava and maize or wheat flour of lower quality in Pakistan, or sorghum in Bangladesh.[3]

From a theoretical point of view, subsidies are the obvious solution to the dilemma between high prices to encourage production and raise incomes of poor food producers, and low prices to safeguard poor food buyers. In practice, the costs of subsidies can be very high. They will depend on the size of the difference between the buying and selling price, and the amount of food to which, or the number of people to whom, the subsidy applies. They will also depend on the costs of marketing if these are borne by state marketing boards. Very high subsidy costs have been incurred, for example, in Egypt where the wheat subsidy alone absorbed 3.5 per cent of GDP in 1980–81 when the domestic price was 72 per cent below the international price. This subsidy was not targeted to specific groups. In 1981–82, the Egyptian government spent 7 per cent of GNP on all food subsidies. Subsidies will increase when they attempt to insulate domestic consumers from generally rising price levels, or rising costs of imports, e.g. as a result of a devaluation.

A third way is to provide income subsidies to the poor. These can take the form of direct subsidies in the form of cash payments or of employment programmes that generate additional incomes for the poor. These can be either public works programmes, or income transfers to the self-employed, or they can take the form of subsidies to private employers, either directly, for employing more men, or indirectly, by subsidies to complementary inputs.

The decline in nutrition from price increases for such income subsidies, however, tends to be greater than the decline in real income because the relative price of food rises and there is some substitution. If the concern is to protect the nutritional level of the poor, targeted subsidies or direct food transfers are preferable. But targeted subsidies are equivalent to income transfers if the quantity of food bought in the absence of the subsidy would be the same or greater, for then the subsidy would only set free income to spend on any other desired goods. Even this qualification needs a qualification, if different members of the household have different propensities to consume food and the direct beneficiary of the subsidised component or of the food transfer under the scheme is someone other than the income earner, such as pregnant or lactating women or school children, or women in cooperatives who benefit directly from the subsidy. It is, however, possible (and likely) that the additional access to food is partly subtracted from the food received within the family.[4]

There is clearly a case for subsidies to food in the transition period, before higher supply prices bring forth the additional food production. The danger is that they become a permanent feature and divert resources from productive investment in agriculture. In this case, a large subsidy programme would relieve present poverty at the expense of increasing future poverty. But a component of any food scheme that reaches the very poor contributes to human capital formation, which is also productive. In practice, many subsidy programmes have assisted the urban middle class rather than the rural poor, and have added to the stream of migrants from the country to the towns.

Subsidies can be combined with rations, so that the very poor are guaranteed a minimum level of food. But food subsidies combined with rations that cover a large population can be very costly. In Sri Lanka, subsidies for food sold in ration shops absorbed 15 per cent of the budget in 1975, and only when the rations were confined to a smaller section of the population in the second half of the seventies did the fiscal costs fall to 7 per cent in 1983. Rationing is also very difficult in sparsely populated areas, where the rural poor often live. The Fair Price Shop Programme in India has served the urban poor better than the rural poor. On the other hand, if the poor reside in distant and isolated areas, it is possible to distribute unrationed food at a subsidized price in local shops. This has been tried in the north-east of Brazil. This system works as long as the subsidy is not so great that it pays to buy the food and resell it in the higher priced markets. If specific groups are selected to be entitled to rations at subsidised prices, there is the additional problem that the same people did not remain poor through time. The Sri Lanka Food Stamp Programme has found it difficult to be flexible with respect to eligibility, as some households shrank in size or enjoyed higher earnings while others sank into poverty. Another drawback of rationing is that it transforms economic goods into economic goodies, dished out to favoured groups.

Rationing through queueing for subsidized food is also often practised. If the alternative use of time by the poor is idleness, this can be a useful system, though it would be easy for the rich to hire unemployed poor people to queue up for them. However, if the poor miss work and earning opportunities, the costs may be too high.

As we have seen in the case of dual markets, the basic difficulty with targeted programmes is that they should reach *all* the poor and *only* the poor, and that is extremely difficult. Whether the subsidies apply to income groups or to types of goods, there are bound to be leakages to non-poor and deficiencies in covering all poor. And though fiscal costs fall as the target population is reduced, administrative costs in hitting the target rise, and overtake the savings at some point.

Those responsible for economies in spending government revenue have, rightly, emphasized the leakages to the non-poor of broadly based schemes such as general food subsidies. But the dangers of the opposite leakage, that of omitting some members of vulnerable groups, are at least equally great for more narrowly 'targeted' schemes. Since the broader schemes' leakages imply that some non-poor are made better off, they also increase taxable capacity, so that leakages of food subsidies to non-poor can be corrected by, for example, taxing tobacco consumption.

It is particularly difficult to reach poor food consumers in the rural areas and the benefits of the subsidies will be concentrated on urban populations. In Bangladesh, for example, two-thirds of subsidized grain were distributed in towns, while most poor people live in the country. If the proportion of poor in the population is large, or if the food staples are largely consumed by the target population, total coverage can more easily be achieved, and costs need not be excessive. There is likely to be a trade-off between complete coverage and costs, and between fiscal costs and administrative costs. A compromise has to be struck in the light of the number of poor and the budgetary and administrative constraints.

An additional difficulty is that there is likely to be a conflict between effective, targeted programmes and political support. The powerful and vocal interest groups, as well as the administering bureaucracy itself, will be prone to support relatively ineffective general programmes, while the neediest groups have less power and influence. The administration of ministering to their needs is not in the interest of the bureaucrats.

In spite of all these difficulties in attempting to combine higher producer prices with lower prices to poor consumers, experience has shown that targeted programmes can be quite effective, as long as the number of poor is not too large, say less than 30 per cent of the population. But we may have to accept a trade-off between higher total food production in response to higher food prices and better nutritional standards of poor food buyers, safeguarded by lower food prices.

14 Small Farmers and Employment

Food subsidies discussed in the previous section are an important short-run measure to alleviate the impact of raising food prices for producers. But they can be very expensive and may make, by themselves, only a limited contribution to a long-term solution. For this, it is necessary to raise the earning capacity of the poor, either by generating employment opportunities or by enabling them to grow more productively the food they themselves consume. These longer term programmes are administratively more difficult, and often run into political obstacles. Experience in India and Bangladesh has shown that the provision of credit to small farmers can be a highly productive investment and that there are few defaults.

The impact of food prices on employment is highly complex, and a full analysis cannot be given here. If the income terms of trade are changed in favour of agriculture and against industry, the impact will depend on the employment intensity of the expenditure of the two sectors. Within the agricultural sector, it will depend on the employment-intensity of the expenditure of the income earners and of different crops. This is, for example, higher for jute than for rice, higher for vegetables than for wheat, and higher for groundnuts than for cotton. A switch from rice to jute in Bangladesh may therefore do more for fighting hunger than a Malthusian attempt to grow more rice. The types of inputs used also affect employment. Mechanization, on the one hand, reduces the demand for labour for a crop but, by making multiple cropping possible, may increase it. Fertilizer may reduce the demand for labour if it makes possible the production of a given output with a lower input of labour per unit of output.

Employment linkages can be traced through production and through consumption linkages, or through foreign trade, or through savings and investment. Each of these, in turn, may be forward or backward. Available spotty evidence suggests that these employment linkages tend to be rather small, but clearly some are larger than others. The International Food Policy Research Institute (IFPRI), for example, is now examining the hypothesis that the bulk of the

71

expenditure of medium-sized farmers tends to be on rural services, such as construction, transport, hotels, restaurants, entertainment, health, education, housing, distributive trades and personal services, that these services are labour-intensive, giving rise to remunerative earning opportunities of underemployed rural labourers, and that the medium-sized farmers also tend to supply more of the food which will be demanded by the newly employed service workers. If this hypothesis proves correct, the implication is that incentives should be extended to the middle-sized farmers, whose expenditure would raise income and nutrition levels of the poorest rural groups. But it is not clear why expenditure on services in the neighbouring region is better than more widely dispersed expenditure that would benefit the poor in other regions.

Another objection to the IFPRI approach is that middle-income farmers may spend their incomes on other items than employment-creating local services. They may save a higher proportion or they may import goods. Since any expenditure on food by both small and large farmers involves a non-agricultural component, such as processing, marketing, transport, etc., part of the expenditure on food is on such non-food items. Evidence from the Philippines shows that the propensities to save and import of middle-income farmers are quite high, and that, on the other hand, local consumption linkages are quite high for small farmers.[1] Backward and forward production linkages are also greater for smaller farmers, because they tend to use more locally produced and repaired implements and process locally on a small scale. Moreover, targeting incentives by size of farm is impossible in practice. The administrative resources would be substantial and, even if available, be better employed elsewhere.

The effects of the various agriculture–industry linkages will depend on technology and employment creation. An improvement in the agricultural terms of trade can benefit industry indirectly, by raising the supply of food, an important wage good, and raising the agricultural demand for industrial products. Equally, an improvement in the industrial terms of trade can benefit agriculture by raising industrial inputs into agriculture, lowering their costs, and increasing demand for agriculture products. The success of these price policies will depend on the adoption of technological change in agriculture, and the growth of employment in industry and the rural sector.

For some purposes the distinction between agriculture and industry is not helpful. Advances are needed on both fronts, for example through well-selected agro-industrial projects. Where conflicts arise, they can often be resolved by the location of industry in rural areas, which makes it possible to make use of seasonably underemployed labour and has the added advantage of saving in the high costs of urban infrastructure. Some of the success stories such as Taiwan have relied on this type of rural industrialization.

15 Rural–Urban Migration[1]

The rapid rate of rural–urban migration is the result of several factors. First, the rate of growth of the total labour force has been high, and remunerative opportunities in the rural sector have been limited. Second, perceived income and security differentials in the two sectors have been high. Third, productivity in food production is often very low as a result of the poor quality of natural resources and low investment in infrastructure, irrigation and technology. This pushes people away from the land. Fourth, there is the pull from the rapid growth of the urban sector and urban services, partly induced by foreign capital and aid. There are also other urban attractions besides higher wages, such as government subsidies to food. The migration can add to labour scarcities on the land, to the high costs of providing urban services, and to a change in demand from domestic to imported goods that cost foreign exchange. Underpricing of agricultural products has contributed to the flow of migrants away from the land. It has been estimated that in Egypt price policies for wheat, rice, maize and cotton have reduced employment by about 5 per cent of the rural population.[2] If higher prices for food and other agricultural products were to reduce the number of urban poor by reducing migration, this would also reduce the target group of poor food consumers for whom food prices have to be kept low.

Measures that accelerate agricultural growth and improvement of rural facilities, and create more rural employment will tend to reduce migration to the towns, unless the turbulence created by change in the countryside stirs up aspirations and the capacity to migrate to superior urban opportunities. Two counteracting forces are at work: first, the income (and security) differential between countryside and town and, second, the financial cost of migrating. In the absence of perfect capital markets, so that potential migrants cannot borrow to finance their moves, a rise in rural incomes will reduce incentives, while raising financial ability. One might *a priori* expect higher rural incomes to raise the migration of the poorest, previously prevented by lack of finance, and reduce that of the better off. But there is some evidence that improved rural employment and living conditions reduce the migration of the poorest, and speed up the migration of the better educated and better off.

Policy-makers have an interest in retaining people in the countryside and on the farms, mainly in order to increase agricultural production and improve rural amenities. Migration to the towns increases the urban population that has to be fed, and swells the ranks of the urban job-seekers, thereby depressing urban wages and incomes from self-employment (but possibly raising urban profits). Agricultural activities tend to be more labour-intensive than industrial ones, and if more people are retained in agriculture the resulting total employment will also be higher. Investment in agriculture also tends to show higher returns than in industry. Both the increase in jobs and the higher agricultural and rural incomes will tend to reduce migration and will therefore raise wages and incomes from self-employment in the towns. Care must, however, be taken that the higher urban organized sector wages are not reflected in higher prices of industrial goods, for then the rural–urban terms of trade improvement for agriculture would be removed again.

It is possible to use different methods to bring agricultural incomes more in line with expected, though often not realized, urban ones. South Korea subsidized agricultural prices (like Great Britain before her entry into the European Common Market). A second option is to use rural public works, such as Maharashtra in India has done. A third option is to encourage the rural location of industry and services, so that new earning opportunities arise for the rural population. As Tibor Scitovsky has pointed out, in Taiwan this had the advantage of saving on budgetary costs of subsidies, and on the expensive urban infrastructure such as housing, schools, urban transport, shopping facilities, etc.[3] Farm families constitute nearly 30 per cent of the population, but only 20 per cent of the labour force is employed on farms. Commuting on a full-time, part-time or seasonal basis to manufacturing, teaching and administrative jobs is facilitated by short distances, good roads, good public transport, and the widespread possession of motor bicycles. As a result, farm incomes are about the same as urban incomes, an important contribution to reduced inequality in the country. Rural–urban migration will also tend to be reduced by raising the security of rural incomes and by reducing rural relative deprivation.

What ultimately matters are, of course, human beings and families, not averages of large sectors, such as rural or urban populations. It is always possible to narrow the differentials between groups, while raising them within these groups. Average rural incomes may then approximate more nearly average urban incomes, while incomes within the rural population become less equal. The conflict is then not one between growth and equity, but between one type of equity and another. Something like this may be happening now in the People's Republic of China, where some rural incomes are now higher than some urban incomes, but inequalities in the countryside have grown.

16 The Politics of Food Prices

The politics of food pricing is concerned with two sets of questions: first, what interests and pressures lie behind current policies of market intervention and the generation of political resources, particularly those that do not appear sensible from an economic point of view; second, what changes in political pressures can lead to reform?

Many governments have both a direct and an indirect interest in keeping food prices low for urban, industrial workers, government officials, members of the armed forces and students, and generating and transferring resources to them from the rural sector. The direct interest is that, as major employers of these groups, the way to keep their real incomes up while resisting demands for higher money wages and salaries, and to raise profits or reduce losses is to keep food prices down. Private industrialists are also their allies. The indirect interest is that these groups can cause a good deal of trouble by rioting and even overthrowing the government when food prices rise. Since governments also have an interest in raising agricultural production, they subsidize inputs into agriculture (fertilizer, machinery, seeds, credit, land), benefiting often mainly large and powerful farmers whose political support they thereby recruit. In countries where such large farmers figure prominently, such as the Ivory Coast and Kenya, agricultural policy takes on a different form.

Governments also have an interest in keeping the exchange rate overvalued (one of the principal tools of depressing the incomes of small farmers) because the resulting need for rationing and allocating foreign exchange gives politically established groups extra money and power. The power of these groups is one of the strongest sources of resistance to changing exchange rates in many countries. Rents caused by economically inefficient interventions present political resources which can be used to organize political support.

Farming interests are opposed by urban workers who want low-priced food; urban industrialists, who want low wages and low prices for raw materials; bureaucrats and white collar workers who want higher salaries and lower food prices; and politicians who run governments which need taxes and which are major employers and industrialists in their own right.[1]

77

Mancur Olson's approach is also illuminating. Agriculture and the rural sector is exploited in low-income countries because the large and dispersed members of this sector can neither organize themselves adequately nor exercise sufficient pressure on the government to act on their behalf. Any one member knows that the benefits from collective action go to everyone, whether or not he had borne the costs of the lobbying or the joint action. Everyone tries to be a 'free rider' and lacks the incentive to organize or pressurize. In advanced, industrial countries this large and dispersed sector are the urban, industrial interests, which are exploited for the benefit of more readily organized, smaller agriculture. Mancur Olson suggests that a more detailed investigation of subsectors, such as particular industrial interests or agricultural interests would throw additional light on this issue.[2]

The analysis of political interest groupings is useful to explain policies with respect to food prices, the prices of inputs, and the foreign exchange rate, but it can be overdone if narrow, competitive self-interest is regarded as the only motivating force. Even if such action were all pervasive, a self-interested distortion added to a system in which others have established distortions can improve the system. Moreover, coalitions to improve the efficient allocation of resources promise gains to all groups if compensation to losers can be incorporated, and should therefore not be excluded. Without going as far as Coase's theorem,[3] according to which, in the absence of transaction costs and with optimizing behaviour, it will always pay either an injurer to compensate his injured victims for accepting the injury, or the injured victims to compensate the injurer for forgoing it, it should not be assumed that the outcomes are necessarily of the prisoners' dilemma type, in which the uncoordinated pursuit of self-interest impoverishes everybody. Real world outcomes will tend to be in between these two extremes.[4] Neither will self-interested, politically competitive behaviour inevitably lead to inefficient outcomes and mutual impoverishment, nor will free negotiations and enforcement of contracts always lead to the most efficient allocation of resources. On the assumptions of Robert Bates, disinterested government would be impossible. On the assumptions of Coase, government would be unnecessary.[5] But we know that, while there are elements of truth in both views, government in the public interest is possible and does occur.

Governments do sometimes transcend their individual and group interests and act in the common interest or in the interest of the poor and weak. The picture of the state painted by Robert Bates as an instrument of ruthlessly amassing wealth and power by those in office, and those on whose support they depend, without regard to either efficiency or the public interest, is surely limited. The principle of mutual impoverishment by competitive, short-term, self-interested political action of interest and pressure groups that attempt to frustrate the working of the invisible hand by rent-seeking and directly unproductive profit-seeking activities has been aptly called the 'invisible foot' by Stephen Magee.[6] It refers to the organized activities of individuals or groups to protect themselves against the working of market forces. The policy problem consists in finding ways to prevent the invisible foot trampling on and destroying the benefits bestowed by the invisible hand.

It is odd that just at a time when writers such as Albert Hirschman, Amartya Sen, Thomas Schelling, Tibor Scitovsky and Harvey Leibenstein deepen our understanding of human choice beyond the simple preference ordering of economic man, some public choice theorists are applying a rather narrow interpretation of selfish economic man to transactions in the political arena. Just as we are getting rid of 'rational fools' in economics, they reappear in political science. It is also odd that writers in the liberal conservative tradition (in the Manchester sense) take a view of the state that equates it to a pure instrument of inefficient exploitation, whereas writers in the Marxian tradition interpret it in much broader terms.[7]

According to one theory of the state, an idealistic and well-informed government, like Platonic guardians, reigns above the interest conflicts and promotes the common good. In this theory the government can do no wrong. The opposite theory, represented by some Chicago economists, and members of the public choice school, holds that the government can do no right. Any intervention by this predatory state with the magic of the market place is bound to make matters worse. Government intervention is not the solution, it is the problem. A third theory, propounded by Anthony Downs, holds that politicians maximize their own welfare by selling policies for votes. The social contract theories say that citizens surrender some of their rights in return for protection and other services from the state. Marxist theory says that the government is the executive committee of the ruling class and always serves the economic interest of that class. But this is open to different interpretations. Some regard the state as acting in the interest of international capital, extracting surplus from the periphery for the benefit of the centre. Others regard it as acting in the interest of an indigenous capitalist class, sometimes against the interest of the capitalists at the centre. On both these views the state acts on behalf of the interest of a ruling class. A more sophisticated version of this theory holds that it is the function of the state to reconcile differences of interest within the ruling class so as to maintain its power and the capitalist mode of production. On this version it is possible to impose price controls on food to keep workers quiet, in spite of the loss of short-term profits that this involves: the system is saved from revolt.

It has been argued in this book that governments are neither monolithic nor impervious to outside pressures, and that the obstacle to 'correct' policy-making is neither stupidity nor cupidity, neither solely ignorance nor solely political constraints. At the same time, there are large areas in which a better analysis and a clearer sense of direction would help, just as there are areas where it is fairly clear what should be done, but vested interests prevent it from happening.

Yet, it is futile (or tautological) to say, in such situations, that the political will is lacking. One does not have to be a behaviourist to believe that behaviour is the manifestation of the will. If the will to action is lacking, there is no point in asking for the will to have the will to action. It only leads to an infinite regress. It is a case of *ignotum per ignotius*. Political will itself can be subjected to analysis, to pressures, and to mobilization, and it is more fruitful to think in terms of the construction of a political base for reform. Let us assume, for the purpose of this chapter, that the correct policies are known and that it is only a question of implementation. Compared with the large and growing literature on rent-seeking and directly unproductive profit-seeking activities, relatively little research has been done by political scientists into the question of how to build constituencies for reform, how to shape reformist coalitions or alliances between groups whose interests can be harnessed to the cause of reform. An exploration along these lines will draw attention to the desirability and feasibility of compensating losing groups, or to mobilization of the power of dispersed, weak or inarticulate groups, or to participatory forms of organization, or to the use of splits within the ruling groups that can be used for the benefit of the poor. It would analyse ways in which the work of the invisible foot can be coordinated to that of the invisible hand, how constituencies can be mobilized for the efficient and equitable allocation of resources.

People are, on the whole, quite good at discovering and pursuing their self-interest, and not much teaching and preaching is needed to steer them along this path. Yet, there are many situations in which the individual is helpless and where procedures, rules or institutions are needed to give expession to his interests.[8]

The rural and the urban poor are weak and powerless, particularly if not organized among themselves. It has been argued in a different context above that in societies in which power and wealth are very unequally distributed, both low producer prices and high producer prices can reinforce the strength of powerful and rich groups. The interests behind low prices are the urban middle class, including the bureaucracy, the military, the police and politically active students, who have access to fair price shops and to imported food aid, and enjoy the benefits of low food prices. The rich and powerful farmers are often protected by subsidies and special allocations of scarce inputs. If this regime is changed to one of higher food prices, again the rich farmers may benefit, while the urban middle class may be protected by special measures under the pretext of protecting poor food consumers.

This does not mean that the interests of urban industrialists always prevail. The location of firms in politically important regions, even if less efficient, or the overmanning of factories to reduce urban unemployment, show that governments often sacrifice industrial profits for the benefit of other interest groups. But the poor rural producers' interests are neglected.

The fact that policies reflect power distributions is evident from the policies of rich countries which tend to tax the relatively poorer urban consumer for the benefit of the frequently better off farmer, while the opposite is true for most poor countries.[9]

The urban bias and discrimination against the rural sector that we observe so widely in the developing countries today were also practised by England and other European countries in the eighteenth and nineteenth centuries, long before current theories of industrialization had been formulated. Adam Smith noted it in his *The Wealth of Nations*.[10]

It is possible to mobilize some powerful interests for the improvement of the fate of the poor, and to give the poor themselves more power. Thus in nineteenth-century England, before workers had organized themselves into powerful trade unions, or had the vote, the urban industrialists had an interest in low food prices in order to keep wages low, and agitated for the repeal of the Corn Laws, against the pressures of the Tory landlords who claimed that this would ruin British agriculture. It did not. The Tory landlords, on the other hand, under the leadership of the Earl of Shaftsbury, opposed to the urban industrialists, agitated successfully for factory reforms, limiting the working day, getting children and women out of the mines, regulating and humanizing conditions in factories, introducing safety legislation, etc., which the industrialists claimed would ruin British industry.[11] It did not. It was therefore the interest conflict between Tories and Liberals that benefited the poor, without destroying either industry or agriculture. Two main lessons can be learned from this experience. First, differences within the ruling group (in addition or as an alternative to the mobilization of the poor) can be used to benefit the poor. Secondly, there is often scope for unperceived positive-sum games, in which feared losses from improvements in policies do not materialize and almost everybody benefits, even without compensation, as was the case in 19th century England.

We should like to know more about the links between politics and food policies. It would be interesting to relate, for instance, the removal of food subsidies to riots, disturbances or falls of governments. It would be useful to trace the groups that have an interest, direct or indirect, in higher food prices. It would be rewarding to trace the interests that can be mobilized for the measures needed in the period of transition.

In many developing countries the problem is how to mobilize interest groups for more remunerative prices to producers, without harming poor food consumers. In some countries large and rich farmers are aligned with urban interests and against small and poor farmers. In others, they make common cause with all agricultural interests. What are the conditions for this to occur?[12] Not only in the advanced industrialized countries have high prices and other forms of protection for food producers reached a point where political forces have to be mobilized for the reversal of these policies. South Korea, for example, has kept its high producer prices, which in the past fulfilled a useful function, for too long, and has created vested interests in their perpetuation. In such cases, political coalitions for their reduction are needed. Subsidies to farmers tend to get capitalized into higher land values. The purchaser of the land with the inflated value would suffer unfairly if subsidies were reduced, and this adds to the difficulty to reduce or eliminate subsidies.

Making politicians and politics endogenous variables is very attractive from the point of view of a relevant and realistic approach, but it raises a profound methodological difficulty. As long as we assume political forces to be exogenous variables, we can use them as levers to change the system: to preserve it, reform it or change it radically. They then also lend themselves to the construction of utopias, without practical constraints. But once political factors are endogenous, they are determined within the system and advocacy of change becomes impossible. What appear to be obstacles in an analysis confined to economic variables become necessities when political variables are incorporated. If we accept determinism we have no choice. In order to get an Archimedean point from which to lift the system, we need some degrees of freedom within which ranges of policies can be regarded as independently determined.

There are some ways out of this dilemma. Max Corden suggests three.[13] First, persons or groups concerned with the national or social interest, rather than with sectional interests, can, from a variety of motives, form part of the process of political pressure groups, and economic analysis can be used by them as the basis for their campaign. International agencies such as the World Bank may constitute such pressure groups, and the conditionality attached to loans, aiming at the best interest of the receiving country, amounts to pressures that counteract the self-interested actions of the urban élites. It is then the external ever jucier carrots, and even bigger sticks that lead the donkey to the Archimedean point that levers the system out of its position. Second, normative economic analysis can show that the lobbies may not be efficient in pursuing their own interests, since they could all be better off if they followed efficient lines and compensated the losers. But this line of reasoning presents the difficulty that differences over the desirability and acceptability of the division of joint gains can prevent their achievement. Once gains are generally accepted, differences over their distribution are just as divisive as conflicts over the distribution without net gains. Compensation may be regarded as either undesirable, because the losers deserve to lose, or as not feasible, or, if feasible, excessively costly. Third, normative analysis may show the inefficiency of the competitive political lobbying process and suggest changes in institutions to produce better outcomes. In other words, the lobbying process need not be accepted as given but is itself subject to change. Of course, this change can itself have costs and may produce non-optimal results. To this we might add a fourth possibility, namely that the rulers do have the common interest at heart, at least partly, and resist the competitive self-interested pressures (see p. 78).

In this chapter the attempt was made to incorporate political variables into the analysis of the response mechanism of food prices. Such an incorporation can provide a sounder basis for analysing responses to food prices than a narrow focus on producers' supply response and consumers' demand response, the way in which economic analysis is conventionally conducted. In principle, all relevant variables should be incorporated in a full analysis. In earlier chapters we have paid some attention to the institutional arrangements which have an influence on responses. If, for example, marketing boards absorb a large proportion of higher prices, the supply response will be different from what it would be if the whole extra profits were to accrue to the farmer. If women do the growing of food but men receive the money from its sale, and make the decisions about expenditure, again supply responses will be different from what they would be if women themselves made the decisions. Or again the system of land ownership, such as sharecropping, and of tenurial arrangements will determine supply responses. Any analysis that claims to be useful to policy-makers must not confine itself to the textbook variables of demand and supply, and markets, but must investigate the institutional and political channels through which demand and supply are expressed.

17 The Budgetary Constraint

The budgetary constraint is fundamental to the basic dilemma. If it did not exist the solution would be easy. The government would simultaneously keep producer prices high and consumer prices low by a general system of deficiency payments to farmers or general food subsidies to consumers. Such a system is possible only if the food is consumed in processed form (e.g. bread). Otherwise, food will be bought at the low, subsidized prices and resold at the higher producer prices. Even advanced countries have found it difficult to sustain such a system, and no developing country has the capacity for such a large subsidy programme, without sacrificing other important objectives such as new productive investment or operating existing investments. Either the country keeps prices paid to farmers low, with the negative effect on agricultural production (though, as we have seen, much less for total output than for one specific crop), or parastatal organizations buy at higher prices and sell at lower ones. The budgetary burden of such a subsidy programme means that other types of development expenditure are sacrificed. Subsidies to food absorb up to 20 per cent of budgetary expenditure in some developing countries. The solution to the dilemma has therefore to be found against severe budgetary constraints. The dilemma has three, not two, horns. Keep producer prices up, consumer prices down, and minimize government costs.

Even when tax revenue (or foreign exchange or surplus earnings of public marketing boards) is available, the priorities for its expenditure set by political pressures often conflict with those that would be indicated by a desirable food pricing and agricultural investment policy. This is one reason why, for example, counterpart funds collected from the sale of food aid are not used to subsidize domestic farmers and thus offset the disincentive effects of the food aid, or why surplus earnings from marketing corporations are not reinvested in agriculture but used to finance other forms of government expenditure.

If general subsidies were to be paid without equivalent taxation, there would be added inflationary pressures. The incidence of such inflation may undo the objectives pursued by the subsidy policy, by imposing an implicit tax on poor producers and consumers. Even if the

85

incidence of inflation falls mainly on better off groups, it has undesirable side effects.

Calculating the effects on the budget of a price support or subsidy programme is a complex matter. First, demand and supply of the subsidized crop will tend to increase. Second, demand and supply of substitute crops will be reduced. If these had been subsidized before, this would reduce the total subsidy; if taxed, it would increase the burden on the budget because tax revenue would fall. Third, inputs into the crop will rise, into substitutes fall. If the inputs into the subsidized crops were subsidized, this again increases the budgetary burden.

Subsidizing the import of food reduces the price to consumers, but it also discourages domestic food production, lowers the incomes of domestic food producers and imposes a burden on the Treasury in the form of the subsidy. The same objective can be achieved, if the country is a food exporter, by taxing exports, with the advantage of a revenue gain to the Treasury. The transfer of income from producers to consumers is the same as in the case of subsidized imports, but extra revenue is levied, instead of spent. Thailand, for example, taxes rice exports with a depressing effect on the domestic price and production. Argentina levies export duties on wheat, grain, corn, sorghum, oil seeds and edible oils. A study by Lucio Reca shows that the net losses to society of this policy 'accounted for more than 10% to society of the value of production [of the crops] in the early 1950s and for some 4% in the early 1970s'.[1] The high taxation of export grains implied a massive redistribution of income from producers to consumers and the government. The dilemma between export taxes as a contribution to sound public finance and higher producer prices as a contribution to higher agricultural output also reflects the dilemma between short-term stabilization policy, pursued by the IMF, and long-term adjustment policy, pursued by the World Bank. It is obvious that tax rates can be so high as to discourage any production, with the result that tax revenue is zero. Equally, if tax rates are zero, no revenue will be collected. It follows that there is a range in which lower tax rates yield higher total tax revenue, because the yield from higher production more than compensates for the reduction in tax rates. Within this range no conflict arises. But in view of the low elasticities of supply, in most realistic situations this range is very narrow.

The policy-maker has to choose between the present system of low supply prices combined with selective input subsidies and a system of high supply prices combined with consumer subsidies. Obviously, higher prices with reduced input subsidies reduce budgetary expenditure. But a substitution of direct consumer subsidies for these input subsidies can be extremely expensive. In Bangladesh the total subsidy on domestically procured grain was more than TK 650 million in FY 1978. The subsidy on fertilizers and modern agricultural inputs by the Bangladesh Agricultural Corporation in FY 1978 and 1979 was more than TK 1000 million or 25 to 30 per cent of government development expenditure in agriculture. In 1981/2 the Egyptian government spent more than $2 billion – about 7 per cent of GNP – on subsidies to food. One solution would be higher output prices with lower input subsidies but high land taxation and income tax. It would combine budget relief with equity and efficiency, but normally runs into political obstacles and inhibitions. We are back at the politics of food pricing.

In several places in this book the importance of the budgetary constraint has been emphasized: in the context of combining consumer subsidies with producer incentive prices, or producer subsidies with low food prices for consumers; of compensating farmers for the detrimental impact of food aid or low-cost imports; of permitting public sector action in technology, infrastructure, institution building or extension services to be induced by price policies; or in instituting a crop insurance scheme to prevent poor farmers from falling below a subsistence minimum and encouraging them to adopt risky innovations. Either the revenue required for these schemes is not forthcoming, or, if it is raised, other development objectives, for which it would otherwise have been used, are sacrificed.

18 The Balance of Payments Constraint

Many developing countries suffer from balance of payments difficulties. Some spend much money on imported food and become dependent on it, and in the case of many African countries the prices and the supply of export crops have declined, although the prices of food imports have declined as much as the prices of their exports. Foreign exchange is very important to agriculture, first, because it often buys some of the inputs into agriculture and some of the consumer goods on which farmers spend their income, and, second, because an overvalued exchange rate is an implicit tax on agricultural exports and income. Many authors regard this as the most important instrument by which agricultural growth is discouraged in economies in which agriculture is an important sector. Thus Malcolm Bale concludes a study of five typical developing countries by saying that 'the implicit taxation of agriculture through an overvalued exchange rate appears to be the single most powerful disincentive to agriculture in the five developing countries examined'. (These were Colombia, Jamaica, Nigeria, Pakistan and the Philippines.)[1] The poorest countries are still heavily dependent on the export of agricultural products. We have already discussed the dilemma that may arise when agricultural crops for export compete, in the allocation of scarce resources, with food crops for domestic consumption.

Absolute returns on both export and food crops should, in many cases, be raised. If the relative rise in the price of food crops is greater, this will improve the domestic food supply, but it may be less labour-intensive and therefore not raise the incomes of the poor as much as the export crop (rice versus jute in Bangladesh), or it may reduce export earnings for agricultural inputs and industrial employment, and the consumer goods on which the farmers spend their income. If the higher food prices lead to the adoption of new techniques total agricultural output will rise and the choice between food crops and export crops is less stark. There have been cases where even in the short run additional export crops did not reduce the supply of food crops, because they use surplus land and labour. Where there is a reduction in food crops, additional imports can temporarily prevent a large increase in food prices. For the longer run, it is important to ensure that farmers get the inputs to grow more food, and that incentives to supply food to the domestic market remain adequate.

In much of the literature there is a strong emphasis on removing overvalued exchange rates as the main obstacle to higher agricultural production for both domestic consumption and exports. Overvaluation of the exchange rate reduces the domestic price farmers receive for their production and exports, and the encouragement to imported duty-free foodstuffs further discourages domestic production.

Brazil implicitly taxed exports at a rate of 22 to 27 per cent between 1954 and 1966 as a result of an overvalued exchange rate. The Egyptian price of rice, already lower than the border price at the official exchange rate, is reduced by another fifth if allowance is made for the overvaluation of the exchange rate. The Mexican peso has for some time been greatly overvalued, discriminating against agricultural production.

The process can be cumulative. If food is subsidised and imports rise, and if the subsidies are financed by a budget deficit, inflation will be accelerated. This will add to the demand for imports, the pressure on the exchange rate, and its overvaluation if it remains fixed. The demand for imported food will further increase and the supply of domestic food be discouraged. If no finance is available, the budget deficit will have to be cut or the exchange rate devalued. But there may be revenue from oil or some other high-priced primary export good.

Paradoxically, the large influx of foreign exchange, whether from oil export revenue or some other primary export product, is often harmful to the agricultural sector. In addition to the disincentive effect of the overvalued currency, there will be a tendency for resources to move out of agriculture into the expanding sector. This does not mean that all the people leaving agriculture will actually find employment in the booming sector. They may simply add to the unemployed. Since the expanding sector usually requires complementary non-tradable services, such as construction, supply limitations here will add further to the inflation, to the overvaluation of the exchange rate which can continue because of the high earnings from the booming sector, and to the damage done to agriculture. Higher domestic incomes from the booming sector lead to extra expenditure on food, but if the prices of agricultural products are determined by world market conditions, it will lead only to extra imports. Agricultural exports will be discouraged, because the prices of non-tradable home goods have risen relatively to those of exports. Where land and other resources compete

for the two uses, the prices of these factors of production will rise and
add to the overvaluation of the exchange rate.

Any discussion of exchange rate policy has to answer three
questions. First, can the real exchange rate be affected by policy
measures? Experience has shown that in some situations a nominal
devaluation is followed by domestic inflation which more or less
quickly restores the old value of the exchange rate or even a higher
degree of overvaluation. This was the case in some African countries,
such as Kenya in 1981, and, to a lesser extent, Mauritius, Madagascar
and Somalia. The mechanism is either consequential money wage
increases, in an attempt to restore real wages, or cumulative price
increases, resulting from the higher costs of imported inputs into
production. If it is inflation above the world level that has caused the
overvaluation and the negation of the devaluation, it is the inflation
that has to be tackled, not the exchange rate.

The second question is, assuming a devaluation is effective, so that
the real exchange rate declines, does it improve the balance of
payments? Devaluation can be counter-productive by increasing,
rather than reducing, the balance of payments deficit. The notion of
the *J*-curve has been constructed to deal with the case where a short-
run deterioration is followed by a longer run improvement. The
question then is how long do we have to wait for the long run to take
effect, and how to finance the interim period.

The third question is, assuming a devaluation is effective, and
improves the balance of payments, what are the costs in terms of side-
effects on other policy objectives, and could the same inprovement
have been achieved at lower costs? Among these side-effects are the
impact on vulnerable groups in the community, particularly children,
women and the poor; the impact on productive investment and long-
term growth; some would pay attention to the impact on relative
income distribution; on employment and on certain regions in the
economy, etc.

The removal of an overvalued exchange rate can have merits, quite
apart from that of stimulating agricultural production. Both the
revenue from export taxes and from import duties can be raised (or
subsidies to imports be reduced) and public revenue be increased,
even without any supply response. The opportunities for rent-seeking
that arise from direct import controls are reduced. Income distribution
may be improved, even without any supply response, if the gainers
from higher domestic prices are small and poor farmers, and, more
generally, if tradables are more labour-intensive than non-tradables.

Higher earnings may be a source of capital for rural investment in lines other than those of which the prices have risen.

No doubt, overvaluing the exchange rate has often been a serious impediment to the advance of agriculture – but its importance can be overstated. Overvaluing the importance of a single element such as the exchange rate, in a complex system of interdependent variables, can be worse than overvaluing the exchange rate. To make a devaluation effective, a host of non-price measures are necessary. As we have seen, there must be roads and vehicles to bring the commodities to the market, and to bring inputs and consumer goods to the farmers; there must be marketing institutions to link ports and markets to farmgates; there must be credit institutions to bring finance to the farmers; there must be controlled and reliable water; there must be research and technology and available information to enable farmers to expand their supply in response to the higher prices, etc. If some of these other conditions are not present, a devaluation may not do any good, and if they are present, it may not be necessary, or only by a smaller amount. The relative importance of these different measures varies between countries and times and is a subject for further research.

Kevin M. Cleaver compares thirty-one countries whose currency has depreciated with others whose currency has appreciated between 1970 and 1981, and their agricultural growth rates. While there is some correlation between higher agricultural growth and currency depreciation, many other factors also enter.[2] In accounting for poor agricultural performance, in addition to overvaluation of the exchange rate,

> other factors such as inefficient government involvement in farm input supply and marketing, population growth, the effort made by government in operating and maintaining agricultural investments, resource endowment, the efficiency of agricultural research, extension, and credit services, politics and other as yet unidentified factors are of much greater importance in determining agricultural growth.[3]

Here again, problems of the transition arise. Assume a devaluation, combined with disinflationary fiscal and monetary policies. If other conditions are right agriculture will expand, but it may expand slowly, whereas the previous protected manufacturing sector may contract rapidly. If world prices of agricultural products fall at the same time, the dislocation will be aggravated. There are important implications for the method used for the adjustment. If a devaluation is indicated, should it be sudden and large, or should it be in a series of small steps? A series of small changes makes for easier adjustments in the context of a growing economy, but also creates or strengthens vested interests against further change. A sudden and large devaluation may inflict more hardships but also has the advantage of combating and preventing the rise of vested interests.

The production of food and other agricultural commodities requires a stable environment, so that the correct investment decision can be made. But the financial and monetary environment has become much less stable than it was before 1970. International capital movements, perhaps a hundred times as large as movements of trade in goods, dominate the determination of exchange rates, and exchange rate fluctuations affect the allocation of resources to agriculture. These two new facts, the growth of international flows of capital and the system of flexible exchange rates, have fundamentally altered some of the issues concerned with food and agriculture: they have altered the feasibility of international schemes to stabilize commodity prices; they have altered methods of project appraisal that use international prices as benchmarks; and they have altered the criteria for allocating resources to agriculture and within agriculture. The solution of these problems will depend upon whether we can make the transition to an orderly world monetary system, or whether the emergence of the USA as the greatest debtor country and the sudden decline in the value of the dollar will cause chaos not only in commodity markets and agriculture, but in the world economy as a whole.

19 International Implications

International interdependence has become a slogan. It is certainly true that, as a result of the technological revolutions in transport and communications, world interdependence has greatly increased. According to the US Presidential Commission on World Hunger 'world grain trade is so tightly integrated that a crop failure anywhere in the world may adversely affect the most remote villages . . . The lives of the estimated 800 million to be living in absolute poverty are put at risk every time there is a drought in Kansas, or floods in India, or a late frost in the Ukraine.'[1]

In one sense, interdependence has clearly increased. When the USA imposed a grain embargo on the Soviet Union in 1980, the Russians substituted purchases from Argentina for those from America. Chile, Peru, Japan, Spain and Italy, no longer able to buy from Argentina, shifted their purchases to the USA. Soviet grain imports reached record levels during the fifteen months of the embargo and Soviet livestock herds continued to expand. This was true in spite of the fact that the USA then supplied more than 40 per cent of wheat and nearly two-thirds of coarse grains traded internationally.

Interdependence in food (as in international trade and in culture) is only partial. The large rural hinterlands, in which the masses of poor people live, are not fully integrated into the system of world trade. Small farmers growing food for their own consumption and rural workers rely largely on local markets, are only partly integrated into national markets and even less so into international markets.

The present system of international relations is not helpful to attempts by the developing countries to improve their food situation. National policies by advanced industrial countries to protect their agriculture and to stabilize prices, understandable though they are, simply cast the problem on others outside their frontiers. Efforts to protect domestic farmers reduce the demand for agricultural exports of the developing countries (though they also probably reduce world prices for food importers), and efforts to stabilize prices in domestic markets destabilize world prices. Some developing countries benefit from lower priced food imports and from food aid (though these benefits do not often accrue to the neediest, but to the urban middle class), but these are short-term advantages to be set against the longer term damage to raising food production.

We therefore live in a world in which the industrialized countries accumulate ever larger food surpluses, financed by growing subsidies, and protected against agricultural imports from the developing countries, while some regions of the world, such as the Soviet bloc, parts of west Asia, and parts of sub-Saharan Africa suffer from food insufficiency.

A more rational system of international cooperation would support the efforts of the developing countries to meet their food needs. If a country intends to switch from a policy of low supply prices for food to one of higher prices, nearer to or above world prices, and combines this with the recommended measures to protect poor food consumers, it creates for itself transitional short-term problems of the kind discussed above. They may take the form of heavy burdens on the budget and on administration, or of political discontent and the threat of riots. If the shift to higher food prices is accompanied by an additional redistribution of income to the poor to improve their access to food, beyond the compensation for the higher prices, there is the possibility of an additional impetus to inflation in the sectors producing goods (including food) on which the income of the poor is spent, accompanied by unemployment in the luxury goods sectors. There may be balance of payments problems caused by additional food imports or capital flight. If the reformist government replaces a repressive dictatorship, previously oppressed groups will assert their claims to higher incomes, with inflationary results. If some groups become disaffected they may organize strikes, sabotage, or even *coups d'état*. All these are familiar troubles for reform-minded governments that wish to change the course of policy. Sir Arthur Lewis once suggested that it would be a useful exercise to issue a manual for reform-minded prime ministers or presidents who, faced with these trials, wish to know what to do. It is hoped that this book, in which the problems of the transition are stressed, is a small contribution to such a manual.

In the critical situation described above, the international community can help in making the transition less painful and disruptive. It can help to overcome an important obstacle to reform – the fear that the cost of the transition to more appropriate policies is too high. It can add flexibility and adaptability to otherwise inert policies set on a damaging course. Adjustment loans have come to be accepted in other contexts, such as the transition to a more rational international trade regime. By an extension of the same principle, adjustment loans should be given to the transition to a more rational food regime. This can take the form of financial or technical assistance to a land reform or a tax reform, or of food aid or a scheme of international food stamps. We have seen that food aid is sometimes criticized for discouraging domestic agriculture. But, as argued above, it would be possible to use the local counterpart funds derived from sales to support farmers growing food, a scheme that would automatically generate the fiscal revenue for price support. This has been tried in Zambia for maize, in Brazil and in Bangladesh for rice. Or the food aid could be used, as it is now, for food-for-work programmes that would generate additional demand and not discourage domestic food production. Food aid could also be provided to ease the transition after a land reform, when agricultural output tends to drop temporarily.

Another way in which donors could contribute to more self-reliant policies is for them to finance the food exports of one developing country to other, food deficit countries. This may apply particularly to Africa, where such a scheme could encourage production and trade on a regional basis.

It is surely more sensible, and capable of mobilizing more political support, to reinforce domestic efforts for creating the conditions in which poor people can earn enough to feed themselves and their families, than to call, as the advocates of a new international economic order have done, for various international schemes whose domestic impact on poverty is quite uncertain. The new international economic order proposed here would support at the global level domestic and regional efforts at self-help.

In any assessment of the role of the international community, one has to evaluate the faults of past international cooperation and assistance, and how these can be avoided in future. Among these are wrong advice on policy, faulty selection of projects, possibly excessive expansion of the service sector (particularly government), etc. If development aid has faults, we should get rid of the faults, not the aid.

20 The Wider Issues of Hunger and Malnutrition

The discussion so far has been mainly concerned with ways of increasing the supply of food and of opportunities for the poor to earn incomes that can buy the food. But the removal of hunger and malnutrition is more complex. It calls for a food strategy. The poor in the developing countries spend about 70 per cent of their total income on food and more than 50 per cent of additional income. Lack of adequate food not only makes people hungry and less able to enjoy life, it also reduces their ability and (by causing apathy) their willingness to work. It also makes them more susceptible to disease by reducing their immunity to infection and other environmental stresses. Prolonged malnutrition among babies and children leads to reduced adult stature; severe malnutrition is associated with decreased brain size and cell number, as well as altered brain chemistry. Malnutrition during pregnancy results in low birth weight, which is a particularly important cause of infant mortality. Children who suffer from severe malnutrition show lags on motor activity, hearing, speech, social and personal behaviour, problem-solving ability, eye–hand coordination and categorization behaviour, even after rehabilitation.

Malnutrition today is not a result of a global shortage of food. Current world production of grain alone could provide everyone with more than 3000 calories (well above an arbitrary average requirement of 2500) and 65 grams of protein daily. And the production of grain in the world is growing more rapidly than population. It has been estimated that 2 per cent of the world's grain output would be sufficient to eliminate malnutrition among the world's 400 to 600 million undernourished.[1] (Note, for example, that 40 per cent of the world's grain is fed to livestock.) Nor is undernutrition primarily a problem of an imbalance between calories and protein. Most village surveys have found that if energy intake is adequate, protein needs are also satisfied. The problem is one of distribution: among countries, regions, and income groups, between sexes and within households. In general, it is the very poor, who spend most of their income on food, who suffer most from undernutrition. In many countries more than 40 per cent of the population suffer from calorie-deficient diets, and about 15 per cent show deficiencies of more than 400 calories per day. Within families it appears that children, and in some societies, such as Bangladesh and northern India, women, particularly when pregnant or lactating, receive inadequate amounts of food. Calorie deficiencies vary by geographical area, season and year. To the extent to which global income distribution cannot be altered, so that food cannot be redistributed from advanced to poor countries, substantially increased production of basic foodstuffs in the developing countries together with a rise in entitlements of the poor must form an important part of the solution. (Food availability per head in the developing countries increased by 6 per cent between 1961–5 and 1979. But this figure conceals wide differences between regions and countries, ranging from 29 per cent in Korea to minus 39 per cent in Ethiopia.)[2] The other part of the solution (in addition to higher food production) is adequate incomes of the poor, including both production for their own consumption and cash to buy food. For farmers this means security of tenure or ownership of land, a regular outlet for sales, and a supply of credit. Extra food production is necessary to meet the additional demand created by population growth and higher incomes per head, and to prevent soaring food prices from cancelling the effects of higher purchasing power. But growth of incomes and of food production, important though it is, is not sufficient to eliminate malnutrition.[3]

Receiving more food does not necessarily meet the basic needs of poor people. It may simply meet the needs of the parasites in their stomachs or of the money-lenders. Malnutrition is a problem of the pathology of the environment, and increasing food intake by itself may not help. Cases have been recorded where is has made things worse, because the extra food consumption of the earning members of families was matched by extra physical efforts, and the rest of the family got less.[4] It may not be food that is needed but education (including changes in tastes and values that cause hunger in spite of adequate incomes), safe water, medical services, or a land reform to permit people to make better use of, and have better access to, the available food supply.

In addition to higher food production and adequate incomes to buy the food, food needs can be met by reducing unnecessary requirements. Food requirements are raised by infections and illness, by long walks to collect firewood or water or other forms of hard work that can be reduced, and by unwanted pregnancies.

Raising the real incomes of the poor so that they can buy more food is clearly one important way of improving nutrition. But this is a slow process, and there are speedier and more direct ways. Iodine deficiency, which can cause goiter, apathy, and proneness to other diseases, is easily remedied by iodizing salt. More difficult to remedy are deficiencies in vitamin A, which can cause blindness and death in children, and in iron, which leads to anaemia and reduced productivity. Protein–energy malnutrition, which may cause irreversible brain damage in children and apathy in adults, is the most difficult to remedy. Yet, it is the most serious problem in undernutrition, followed by deficiencies in iron and vitamin A.

Apart from emergency action in a famine, nutrition policies for the chronically malnourished poor call for a long-term, sustained effort. Intervention can take the form of agricultural policy, supplementary feeding, food fortification programmes, food subsidies and rationing, reducing unnecessary food requirements, and complementary policies in nonfood sectors.

Since the poor and the rich do not spend their money on the same kind of food, policies that encourage greater production of poor people's food – such as cassava, corn, sorghum, and millet – can help reduce malnutrition. Food marketing and storage programmes can reduce regional, seasonal, and annual variations in supplies and prices. Policies to encourage production of food for the poor should extend to all aspects of agricultural policy, including research, extension programmes, credit, and marketing.

Supplementary feeding may take place in schools, at work or at clinics for pregnant or lactating women. With the receipt of extra institutional food, however, meals at home may be curtailed, so that the vulnerable groups do not get much additional food, and, at least in the case of schoolchildren, these programmes do not reach the groups particularly at risk, such as children below school age. Here again, as for the causes of malnutrition, the ease of intervention (because schools already exist and delivery is cheap) is inversely related to its importance. Food supplementation at the work place, if neither the food nor the extra energy is diverted to other activities, serves both a basic need and productivity. The most vulnerable groups, such as children under five, large families and landless households are also most difficult to reach; the time when needs are greatest, the wet season, is also the time when access is most difficult; and the location, the countryside, where most malnutrition exists, is more difficult to reach than the towns.

Special foods and food fortification, as in the case of protein and vitamin fortification and salt iodization, have been successful up to a point, though they meet with both technical and political difficulties.

The problem of eliminating hunger is more complicated than raising food production and reducing population growth. Policies in sectors other than food are essential for better nutrition. Safe water and the prevention of intestinal diseases, a reduction of unwanted pregnancies, and the elimination of unnecessary work would enable people to absorb the same amount of food more effectively. Education, can help people spend their money more wisely, to reduce irrational prejudices and prepare food more economically and hygienically; they can learn to complement their present diet with local food. The battle against early weaning and the use of baby formulas has hit the headlines, but the desire of women to cease breastfeeding is often part of the general process of modernization and the desire to emulate the more advanced groups in the country.

In assessing the extent of malnutrition and undernutrition, we need to know not only the number of malnourished people, but also the severity, i.e. by how much they fall short, the duration of undernutrition for any individual (for it is worse for one to be inadequately fed for four weeks than for four for one week), and the concentration on particularly vulnerable groups, such as children under four and pregnant women.

Malnutrition and undernutrition are the result of a complex set of conditions, all stemming from poverty. Although most people suffering from calorie deficiencies are poor, not all poor people suffer from such deficiencies. Some high-income countries and groups of people suffer from considerable malnutrition and some low-income countries have none. This shows that much can be done at quite low-income levels. There are short cuts and we do not have to wait until incomes have risen.

21 Recommendations

Certain conclusions have emerged from the previous discussion, though much remains inconclusive and in need of further work. (Some of these areas are indicated in the next chapter.) The first point to bear firmly in mind is that adequate supply of food is not enough to ensure the eradication of hunger, malnutrition, undernutrition and starvation. Policies must also ensure that people have the purchasing power or other forms of 'entitlements' (such as access to subsidized or free food rations). Indeed, an integrated nutrition policy must embrace much more than ensuring supply of, demand for, and access to food. It must aim at adequate health standards, particularly the elimination and prevention of intestinal and parasitic diseases, so that the food is properly absorbed by the body; at adequate education, particularly of women, so that people know what food to eat, how to prepare it, and how to keep themselves healthy through hygienic practices; and at ways to improve the distribution of food within the household, so that vulnerable groups, such as pre-school age children or pregnant and lactating mothers, get enough to eat. These changes often depend on changes in the political power structure. The contribution of price policy is to combine incentives to producers with ensuring access to food by consumers. Once the problem of hunger is understood as in need of a multipronged attack, the danger is seen to be, as Amartya Sen has pointed out,[1] not Malthusian pessimism, but Malthusian optimism, i.e. the view that all that is needed is to make the land more fertile and women less fertile: higher prices for food and lower benefits for children. We now know that we can witness hunger in the midst of large food surpluses.

The benchmark for producer prices is world prices, for they represent the opportunity costs of domestic resources. In many countries producer prices have, by a variety of policy measures, been pushed below this benchmark. In those cases discrimination against small and poor agricultural producers should be removed. It is, however, not current world prices but the trend of future prices that should be used as the guideline. They should be calculated in a major convertible currency. Using the trend also eliminates paying heed to temporary fluctuations in prices which are a disincentive to agricultural investment and hurt poor consumers. To the estimated future price trend should be added perhaps 15–20 per cent, as an insurance against world prices rising steeply, developed countries reversing their protectionist policies, or food becoming unavailable in the world market.

In using world prices as a benchmark for domestic producer prices, it is important to ensure that higher prices are not absorbed in marketing margins, either by monopolistic private middlemen or by inefficient or corrupt state marketing boards. Indeed, in many cases a reduction of these margins, by a reform of the distributive and marketing system or their capture by the growers themselves through cooperatives, can bring producer prices nearer to world prices.

The announcement of these prices has to be timely, i.e. before the planting season, so that farmers can incorporate them in their decisions. If the gap between current and world prices is large, a gradual transition is probably easier, but confidence must be established that the higher prices will continue.

Policy must also ensure that agricultural inputs such as fertilizers, seeds, machinery, and transport are available in adequate amounts at the right places. There is no point in raising prices and creating incentives for higher production if the inputs into agricultural growth are not available. There must also be an assurance of markets and the ability to get the crops to the markets. For each of these purposes, namely (i) ensuring demand and access to markets, (ii) not discriminating against agricultural production, and (iii) ensuring inputs, macroeconomic policies relating to exchange rates are at least as important as microeconomic policies for specific groups or crops. An overvalued exchange rate can do considerable harm to food production for domestic consumption and foreign exchange earnings for imported agricultural inputs, though adjustments in the exchange rate by themselves are frequently insufficient.

If a country enjoys a fairly equal distribution of land, has no landless labourers or, if its landless labourers are remuneratively employed in agriculture, rural industry, services or public works, if efficient institutions for credit and marketing exist, if there is an adequate infrastructure, and if the right agricultural technology is available (or employment opportunities exist in labour-intensive exports), price policies to encourage higher output can be effective by themselves. The examples are Japan, Taiwan, the Republic of Korea and Costa Rica.

Where land is unequally distributed and large commercial farms or plantations coexist with small peasants and many underemployed landless, where institutions for credit and marketing are absent or weak or inefficient or corrupt, and where an appropriate technology does not exist, e.g. because there is no irrigation, or inadequate research has gone into the crops suited for the soil, price policies to raise producer prices may be ineffective or counter-productive both for efficiency and equity, unless accompanied by action on the institutional and technological fronts. If these actions are taken, the role of prices is somewhat reduced. If they are not taken, price policies may contribute to inducing the institutional and technological reforms, though evidence for this is weak and controversial. If incentive prices do not include those other reforms, they may still be advisable if the worst sufferers are cushioned against the detrimental impact of higher prices by selective policies that attempt to cover the vulnerable groups. Such policies may aim at subsidizing specific foods, where these are largely consumed by the poor, or at subsidizing vulnerable groups, through income or targeted food subsidies such as food stamps.

There are policies which aim specifically at helping the production and earnings of the poor. One is to target scarce inputs, including land, to the poor. A second is to introduce efficient, labour-intensive methods of production, especially in slack seasons, but to make them not too calorie consuming, so as not to raise unnecessarily food requirements. A third is to produce sources of cheap calories rather than dairy products or vegetables. Unfortunately, some of these may conflict. The cheap calorie-intensive crops may be less labour-intensive than some luxury or export crops.

22 Research Needs

In various places in the book indications have been given where further research is needed. Action should not wait for the results of this work, for the main thrust of policy is clear. But the details and magnitudes could be improved by this work. A list of some of these issues would include:

1. The effect of agricultural prices on innovation, its adoption, adaptation and diffusion.
2. The effects of prices on private and public investment in agriculture, on crop yields and on mechanization.
3. The effect of prices on the pattern of consumption and the consequential impact on employment and foreign exchange. The employment linkages would explore, in addition to the direct employment generated by consumption, forward and backward linkages, linkages through foreign trade, and through the savings/investment nexus.
4. The impact of price policies on income distribution. The links between price policies and initial asset, income and power distributions.
5. The costs and benefits of alternative pricing and control policies.
6. The merits and drawbacks of targeted food subsidies and how to achieve the objective of covering *all* and *only* the poor. Where this can only be approximated, the trade-offs between coverage and costs, and between fiscal and administrative demands. The best way of phasing out subsidies.
7. The relation between food prices and urban wage rates and incomes in the informal sector.
8. The linkages between agricultural and industrial growth. Employment linkages through production and consumption linkages, through foreign trade and through savings and investment, forward and backward.
9. The impact of different price policies on the budget.
10. The relationship between farm size and efficiency, and the factors influencing this relationship.
11. The change in nutritional standards as change occurs from subsistence to cash economy.
12. The impact of food aid on tastes and food consumption patterns.

13. A detailed investigation of what Professor Harvey Leibenstein has called the micro-micro type into the decision-making process of farmers and members of their families, and their time allocation. The distribution of food within the household and the forces influencing it.

14. The scope and the limits of using international prices as benchmarks for setting domestic agricultural prices, and more specifically the limits imposed by fluctuating prices, by uncertain availability of food imports, by the fact that a country may sometimes be a food importer and at other times a food exporter, and by the uncertainty of future international prices.

15. In the field of political science, research is needed on (a) political opposition to higher food prices and (b) the potential reformist coalitions that would enlist the support of powerful groups for reforms.

16. Problems of the transition to a more rational food regime. Who are the main sufferers and, if desired, how could they be cushioned against hardships?

17. The role developed countries and international organizations can play in assisting adjustments to improved food price policies in developing countries.

23 Statistical Requirements

There are numerous good studies of the response of specific crops to the relative increase in the price of that crop, but total agricultural response is less well documented. For this purpose, it would be useful to gather data on the terms of trade between agriculture and industry, and data on quantity supplied over a period.

For the purpose of studying supply responses it would be useful to break down data for the responses of different types of farmers; e.g. small peasants, middle-sized farmers, large commercial farms and plantations, multinational corporations, etc. In this way the aggregated data at present available could be decomposed into the responses of different groups, the size and importance of which changes in the course of development.

In order to determine the impact of price policies on income distribution it would be useful to collect cost-of-living indices for different income groups. If, for example, the price of food rises by more than the general cost-of-living index, the cost of living of the lower deciles, who spend a higher proportion on food, rises by more, and with constant money income, their incomes fall by more. There is a dearth of such data, compared with the attention that has been paid to international comparisons of purchasing power.

It would also be useful if supply responses could be broken down by (a) responses of labour time supplied; (b) responses of specific crops with different labour intensities to higher relative prices; (c) responses of the marketed surplus in relation to changes in consumption retained in the household; (d) responses to the officially marketed surplus compared with non-official sales.

24 Summary and Conclusions

The starting point of this book is the dilemma faced by policy-makers in many developing countries: should the price of food be high in order to stimulate production, or low in order to prevent poor food buyers from starving. There are also many poor. food sellers who would benefit from higher prices. The role of prices and price policy in the light of the objectives of production and equity is then discussed. One conclusion is that in some situations 'getting prices right' by itself, without complementary action in the public sector (on technology, roads, health, extension services, etc.), can be ineffective or counter-productive. In addition to the impact of prices on efficient resource allocation, we discussed in some detail how changes in prices affect different vulnerable groups, and how these can be protected in the transition from a set of wrong policies to better policies. In this context, we made suggestions as to how the international community can help when countries adopt sound food and nutrition policies. A special chapter is devoted to a discussion of the politics of food pricing policies, in which the view is rejected that the Invisible Foot inevitably tramples on the good work of the Invisible Hand, and ways were pointed out in which political constituencies can be built for political reform. The ultimate purpose of these policies is to reduce hunger and undernutrition in the world.

Notes and References

1 The Dilemma

1. Some authors would add labour. Agricultural labour did make, historically, an important contribution to industrialization. But in present conditions in most developing countries the problem is an excess of labour in the agricultural sector. This is the result of the faster rates of population growth, the more capital-intensive techniques used in manufacturing industry, and the higher level of real wages maintained by trade unions and conventions. The problem is normally not that of drawing on rural labour for urban industry, but discovering non-agricultural rural technologies that will absorb the surplus labour productively at remunerative incomes. This is so in spite of occasional labour shortages at harvest time. See the discussion of the various surpluses below.
2. Cathy L. Jabara, 'Agricultural pricing policy in Kenya', *World Development*, vol. 13 (May 1985) no. 5.
3. W. A. Lewis, 'Economic development with unlimited supplies of labour', *The Manchester School*, 1954.

2 Multiplicity of Objectives

1. There have been riots, as a result of rising food prices, in Brazil, Bolivia, Peru, the Dominican Republic, Tunisia, Zambia and Egypt, among others.
2. Clearly there are other constraints also, such as the general scarcity of resources, the available technology, managerial attitudes, markets, political resistance, etc. The above-mentioned are singled out because they can sometimes be treated as independent objectives. The other constraints are discussed in their context.
3. *The Wall Street Journal*, 31 July 1985.

3 Towards a Country and a Crop Typology

1. See Paul Streeten and Diane Elson, *Diversification and Development: The Case of Coffee* (New York: Praeger, 1971).

4 What Are Price Policies?

1. Michael Lipton and Carol Heald, 'The European Community and African food strategies', Centre for European Policy Studies, Working Document, no. 12 (Economic), December 1984.

5 Allocational Efficiency and Higher Food Production

1. An illustration of providing the means without the incentives was the Tanzanian policy that intended to overcome the absence of deliveries to distant rural regions resulting from a pan-territorial pricing policy. The government then provided for the extra transport costs, without making payments contingent on actual delivery to the distant areas.
2. Striking evidence for this is provided by the Stanford Project on the Political Economy of Rice in Asia. C. P. Timmer and W. Falcon, 'The Political Economy of Rice Production and Trade in Asia', in L. G. Reynolds (ed.) *Agriculture in Development Theory* (New Haven: Yale University Press, 1975).
3. Thus the World Bank's Report *Accelerated Development in Sub-Saharan Africa: An Agenda for Action* (Washington, D.C., 1981), after discussing the supply response to the relative price of a crop, devotes a footnote to saying 'the question of aggregate response is more nuanced'. (William Safire wrote that the word 'nuanced' will forever bear the situation room brand of Alexander M. Haig, Jr.) If 'nuanced' means subject to a subtle or slight variation, this statement is quite wrong. The confusion may have arisen because before the mid-sixties farmers were considered as unresponsive to price incentives (or even perversely responsive). The massive evidence in the late sixties showed that they were indeed responsive to allocating resources *between* activities and crops. The impression has sometimes been wrongly applied to *total* response, which is much less, not because farmers do not seek profits but because supply constraints exist.
4. An important exception is a paper by Hans Binswanger, Yair Mundlak, Maw-Cheng Yang and Alan Bowers, 'Estimation of aggregate agricultural supply response', World Bank Discussion Paper, August 1985.
5. *Zambia*: J. Pottier, 'Defunct labour reserve? Mambwe villages in the post-migration economy', *Africa* vol. 53 (1983) no. 2, pp. 2–24.
    Ghana: A. Shepherd, 'Agrarian Change in Northern Ghana: Public Investment, Capitalist Farming and Famine', in J. Heyer *et al.* (eds) *Rural Development in Tropical Africa* (New York: St Martin's Press, 1981).
    Kenya: M. Cowen, 'Food Production in Central Province, Kenya, 1945–1970', in R. Rotberg (ed.) *Imperialism, Colonialism and Hunger in East and Central Africa* (Lexington, D.C.: Heath, 1983).
    Ivory Coast: J. P. Chauveau *et al.*, 'Histoire de riz, histoire d'igname: le cas de la moyenne Cote d'Ivoire', *Africa*, vol. 51 (1981) no. 2, pp. 621–58.
6. See J. Mellor, 'Food price policies and income distribution in low-income countries', *Economic Development and Cultural Change*, vol. 27 (October 1978) no. 1; R. Herdt, 'A disaggregate approach to aggregate study', *American Journal of Agricultural Economics*, vol. 52 (November 1970) no. 4, pp. 512–20; M. Bond, 'Agricultural response to price in sub-Saharan African countries', *IMF Staff Papers*, vol. 30 (November–December 1983) no. 4. For a review of some empirical analyses, see H. Askari and J. T. Cummings, *Agricultural Supply*

Response (New York: Praeger, 1976).

7. Hans Binswanger, Yair Mundlak, Maw-Cheng Yang and Alan Bowers, 'Estimation of Aggregate Agricultural Supply Response', World Bank mimeo, p. 30. For an analysis that attempts to separate the influence of prices and non-price measures, see Ajay Chhibber, 'Long-term agricultural growth in developing countries', FAO September 1983, mimeo.

8. W. L. Peterson, 'International farm prices and the social cost of cheap food policies', *American Journal of Agricultural Economics*, vol. 61 (February 1979) pp. 12–21. Peterson used cross-section (farm household) data for 1962–4 and 1968–70 from fifty-three countries. A loglinear function was estimated with real prices received for all farm products in terms of kilograms of commercial fertilizer that could be purchased with 100 kilograms of wheat equivalents, a weather variable approximated by the long-run average annual precipitation, and a technology variable approximated by the number of research publications for each country.

9. Ajay Chhibber, 'Long-term agricultural growth in developing countries: a framework for comparing the role of price and non-price policies', FAO Discussion Paper, September 1983. When Chhibber introduced an irrigation variable into Peterson's supply function he found that it reduced the elasticity from 1.27 to 0.97. In Appendix 2 of a paper by Hans Binswanger, Yair Mundlak, Maw-Cheng Yang and Alan Bowers, 'Estimation of aggregate agricultural supply response' Agriculture and Rural Development Department of the World Bank, August 1985, the Peterson analysis is also subjected to a critical review. As they add sister variables the supply response becomes actually negative. Also, with more recent price data, they arrive at a much lower supply elasticity.

10. A strong statement of the case for raising agricultural production through better price incentives is to be found in T. W. Schultz (ed.) *Distortions of Agricultural Incentives* (Bloomington: Indiana University Press, 1978).

11. We have argued that, in general, and not only for food and agriculture, the correct way to pursue a strategy of equitable and efficient growth is to combine actions on three fronts: a correct pricing policy, a technology policy, and the creation of the correct institutions, particularly land distribution and tenure arrangements. To this has to be added population policy. Advance with any one prong without the others can be counter-productive. See Frances Stewart and Paul Streeten, 'New strategies for development: poverty, inequality and growth', *Oxford Economic Papers*, vol. 28 (November 1976) no. 3.

12. See Raisuddin Ahmed, 'Agricultural price policies under complex socio-economic and natural constraints: the case of Bangladesh', International Food Policy Research Institute, Research Report, no. 27 (October 1981); and 'Foodgrain supply, distribution, and consumption policies within a dual pricing mechanism: a case study of Bangladesh', International Food Policy Research Institute, Research Report, no. 8 (May 1979).

13. K. N. Raj, 'Agricultural growth in China and India, the role of price and

non-price factors', *Economic and Political Weekly*, 15 January 1983.

14.   Y. Hayami, 'Conditions for the diffusion of agricultural technology: an Asian perspective', *Journal of Economic History*, 34 (March 1974), and S. Ishikawa, *Economic Development in Asian Perspective* (Tokyo: Kinokuiya Bookstore, 1967). The references are in Ajay Chhibber, 'Long-term agricultural growth in developing countries', FAO 1983, mimeo.

15.   Tibor Scitovsky, 'Economic development in Taiwan and South Korea 1965–81', Stanford Food Research Institute, vol. XIX (1985) no. 3, p. 130.

16.   Yujiro Hayami and Vernon W. Ruttan, *Agricultural Development: An International Perspective* (Baltimore, MD: Johns Hopkins University Press, 1971; 2nd revised edition, 1984).

17.   M. Lipton and C. Heald, 'The European Community and African food strategies', Centre for European Policy Studies, Working Document no. 12 (December 1984).

18.   Robert Crown, 'Price policy and supply response in Togo', paper for Seminar on Agricultural Pricing and Trade Policy, EDI, World Bank, Amsterdam (April 1986).

19.   Raj Krishna, 'Some aspects of agricultural growth, price policy and equity', *Food Research Institute Studies in Agricultural Economics, Trade and Development* vol. 18 (1982) no. 3.

20.   For a discussion of the perverse effects of price increases in conditions of non-availability of consumer goods to farmers on output, see p. 31.

21.   See Frances Stewart, 'Alternative conditionality', *World Development* vol. 1, (1984), and *Planning to Meet Basic Needs* (London: Macmillan, 1985) chapter 9.

22.   Peter Bertocci in *Rural Development in Bangladesh and Pakistan*, edited by Robert Stevens, Hamza Alavi and Peter Bertocci (Hawaii, 1976); Geoffrey Wood, 'Rural class formation in Bangladesh, 1940–1980', *Bulletin of Concerned Asian Scholars*, vol. 13 (1981) no. 4; *Contradictions and Distortions in a Rural Economy: The Case of Bangladesh*, (SIDA, 1979).

23.   Rex Mortimer (ed.), *Showcase State: The Illusion of Indonesia's Accelerated Modernisation* (Sydney, 1973).

24.   David Felix 'Development economics and the dynamics of consumption', paper presented at the session on Thirty Years of Development Economics at the Meeting of the Eastern Economic Association, 1985.

25.   The widely used notion of price 'distortion' is not as clear as it may seem. Distortion is the deviation from some natural, proper, legitimate norm. But many would deny that prices determined at the border through international trade, or in free markets under *laissez-faire*, necessarily reflect such a norm. Any one of an infinite number of different income distributions would produce a different set of relative prices. Free market prices also reflect monopoly power and do not reflect externalities in consumption and production. In conditions of widespread unemployment and underemployment wage rates do not reflect the opportunity or scarcity cost of labour. For reasons such as

these, the notion that government interventions 'distort' an otherwise correct set of signals and incentives is highly misleading. In the presence of such 'private' distortions, the addtion of 'public' distortions can be a beneficial corrective. This, however, does not affect the argument that border prices, however 'distorted', represent the opportunity costs for a small country and are useful benchmarks, as long as the qualifications discussed below are borne in mind.

26. See Cathy L. Jabara, Agricultural pricing policy in Kenya', *World Development*, vol. 13, (May 1985) no. 5.
27. Malumba Kamuanga, 'Farm level study of the Rice production systems at the Office de Niger in Mali: an economic analysis', Ph.D. dissertation, Department of Agricultural Economics, Michigan State University, East Lansing, Michigan, 1982, quoted in Carl K. Eicher, 'Facing up to Africa's food crisis', *Foreign Affairs*, Fall 1982.
28. *Accelerated Development in Sub-Saharan Africa: An Agenda for Action* (Washington: World Bank, 1981) p. 56.
29. Howard Dick, 'Survey of recent development', *Bulletin of Indonesian Economic Studies*, vol. XVIII (March 1982), no. 1, p. 30.
30. M. Lipton and C. Heald (1984) quoting G. Schmidt, 'The interaction between the formal and informal marketing system for maize and beans in Kenya', IDS, University of Nairobi, 1978.

7 The Total Picture

1. See Paul Collier, 'Peasant supply response in rationed economics', *World Development*, forthcoming.
2. Will Rogers said that prohibition is better than no liquor at all. In this case there would be no liquor at all.
3. Ramgopal Agarwala, 'Price distortions and growth in developing countries', World Bank Staff Working Paper, no. 575 (1983). The general conclusion of this study, that price distortions explain only a small part of the variance in growth rates of different countries, and that 'prices matter for growth, though not only prices' is entirely in line with the argument of this book.
4. The suggestion was made by Andrew Kamarck at a meeting of an advisory group of the Agricultural Development Council.

8 Aggregate Growth

1. See e.g. B. F. Johnston and J. W. Mellor, 'Agriculture in economic development', *The American Economic Review*, vol. LI (September 1961) no. 4.
2. If industry employs 20 per cent of the labour force, and if the labour force is growing at 3 per cent per annum, extra industrial employment would have to be 15 per cent in order to absorb only the new entrants every year, making no impact on all the existing unemployed. In no low-income country has industrial employment grown by anything like this rate.

3. The price rise will lead to increased output, unless the work/leisure preference reduces the number of hours worked. Out of this larger output more may be retained for own consumption and more sold to the market, unless the demand for own consumption is so large as to reduce the total amount of food sold to the market.
4. See the references on p. 15.
5. *World Development Report*, 1982.

9 National Food Security and Food Aid

1. This section owes much to discussions with Professor Hans Singer. See also his 'The Terms of Trade Controversy and the Evolution of Soft Financing', in Gerald M. Meier and Dudley Seers, *Pioneers in Development* (Oxford: Oxford University Press, 1984); H. W. Singer and Simon Maxwell, 'Food aid to developing countries: a survey', *World Development*, vol. 7 (March 1979) no. 3, pp. 225–47; 'Food aid and development: issues and evidence', jointly with E. J. Clay, World Food Programme Occasional Papers no. 3; 'Use and abuse of local counterpart funds', *International Development Review*, vol. 3. (October 1961) no. 3; *Food Aid Policies and Programmes: A Survey of Studies of Food Aid*, WFP/CFA 5/5–C (Rome: FAO, 1978); *Development Through Food: Twenty Years' Experience* (with S. J. Maxwell), in WPF/Government of the Netherlands, 1983; *Food Aid and Development: The Impact and Effectiveness of Bilateral PL 480 Title I Type Assistance* (Brighton: University of Sussex, IDS, February 1982) (with E. J. Clay); 'Pricing policies for food aid' (with E. J. Clay), *Technical Paper* no. 8, Commonwealth Consultative Meeting on Food Pricing and Marketing Policies for Developing Countries, London, 3–6 May, 1983.
2. Thus in the plentiful year 1970 annual food aid exceeded 12.5 million tons, whereas in the food crisis of 1973–4, when the price of wheat rose by 50 per cent, annual shipments fell to below 6 million tons. Not only the timing but also the country distribution serves the political, economic and military interests of donor countries. Thus in 1982 and 1983 Egypt received 18 per cent of the food aid distributed by the Food Aid Convention.
3. The argument that counterpart funds should be used for deficiency payments to farmers applies also to subsidized food imports, or to those admitted at an overvalued exchange rate.
4. In a chapter, 'Aid to India', in *The Crisis of Indian Planning* (eds Paul Streeten and Michael Lipton, Oxford University Press, 1968), Roger Hill and I tried to calculate the damaging impact of food aid on Indian agriculture.
5. Food aid has, however, increased since 1975. In the sixties it had been as high as 16–17 million tonnes in some years. In 1973–4 the cereal tonnage had fallen to 5.5 million tonnes. In 1976–7 it was 9.0 million tonnes and in 1984–5 had risen to 10.4 million tonnes. The 1985–6 figure will be higher because of emergency aid to sub-Saharan Africa. There has been an increasing proportion of non-cereal food aid, not covered by these

figures, especially EEC aid in dairy products. The aid component of food aid has also increased and more has gone to the poorest countries.
6. Dunstan S. C. Spencer, 'Agricultural research in sub-Saharan Africa: using the lessons of the past to develop a strategy for the future', May 1985, mimeo.

10 Price Stabilization

1. See *Crop Insurance for Agricultural Development, Issues and Experience*, edited by Peter Hazell, Carlos Pomareda and Alberto Valdes (Baltimore: Johns Hopkins University Press, for IFRI, 1985); and Peter Hazell and Alberto Valdes, 'Choosing the right role for crop insurance', *Ceres*, vol. 17 (May–June 1984).
2. Stabilization round an average of fluctuating prices implies a loss of consumers' surplus and a transfer to producers if supply fluctuates, and a loss of producers' surplus and a transfer to consumers if demand fluctuates. Assume parallel shifts of the supply curve (Fig. 10.1) and the demand curve (Fig. 10.2) of equal probability. S_1 stabilizes supply in the middle and D_1 stabilizes demand. Stabilization leads to a loss of consumers' surplus *ABEH* and a gain of *GHEF*. Since the triangles *ELB* and *FME* and the rectangles *AKMH* and *HMFG* are identical, there is a net loss of *KLEM*. Similar reasoning applies to fluctuations in demand, the shaded area showing the net loss. This assumes absence of stabilizing and destabilizing speculation. For a fuller analysis the gains from risk reduction and the costs of the method of stabilization (e.g. buffer stocks) would have to be taken into account. Newbery and Stiglitz (1981: see note 4) have raised valid objections to this approach.

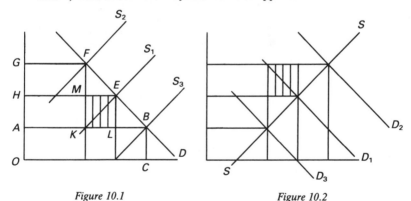

Figure 10.1   *Figure 10.2*

3. If the food is traded internationally, and if the country is small, so that it cannot affect its international terms of trade, the price of the food to the country is given from the outside, and cannot be affected by changes in domestic production. Only the prices of foodstuffs that are not traded internationally are determined by domestic supply and demand.

4. For a full discussion of these issues, see David M. G. Newbery and Joseph E. Stiglitz, *The Theory of Commodity Price Stabilization* (Oxford: Clarendon Press, 1981). They focus on stabilization of consumption rather than income, thereby permitting credit markets to function as a substitute for price and income stabilization.

5. The popularity of these methods among the developing countries and with the United Nations Conference on Trade and Development is largely due to the hope of combining them with restrictive agreements that not only stabilize but also raise the prices of the primary products exported by the developing countries. This, of course, raises quite different questions, not discussed here.

6. 'The world food and hunger problem: changing perspectives and possibilities, 1974–1984: An independent assessment presented to the World Food Council', WFC/1984/,1984. Walter P. Falcon, C. T. Kurien, Fernando Monckeberg, Achola P. Okeyo, S. O. Olayide, Ferenc Rabar and Wouter Tims, p. 29.

7. 'The major result of our analysis is to question seriously the desirability of price stabilization schemes, both from the point of view of the producer and of the consumer.' David M. G. Newbery and Joseph E. Stiglitz, *The Theory of Commodity Price Stabilization: A Study in the Economics of Risk* (Oxford: Clarendon Press, 1981) p. 23.

8. See Nicholas Kaldor, 'Inflation and recession in the world economy', *Economic Journal*, 1976, and 'The role of commodity prices in economic recovery', *Lloyds Bank Review*, July 1983, no. 149. See also the work of P. Sylos Labini, on which Kaldor's second article draws.

9. Kaldor (1983), p. 30. See also S. M. Ravi Kanbur and David Vines, 'North–South interaction and commod control', Centre for Economic Policy Research, Discussion Paper no. 8, March 1984. If, however, Peter Drucker is correct in thinking that the link between primary products and manufacturing has, largely as a result of technological progress, been broken or greatly weakened, the Kaldor case for price stabilization is also weakened. See Peter Drucker, 'The changed world economy', *Foreign Affairs*, Spring 1986.

11 Export Crops *v.* Food for Domestic Consumption

1. See Barbara H. Chasin and Richard W. Franke, letter to the *New York Times*, 25 December 1984.

2. In private correspondence.

3. *Basic Needs in Danger*, a Report of the Jobs and Skills Programme for Africa of the International Labour Organization, 1982.

4. Michael Lipton and Carol Heald, 'The European Community and African food strategies', Centre for European Policy Studies Working Documents, no. 12 (1984) pp. 25–26.

12 Income Distribution and Poverty

1. John Mellor, 'Food price policy and income distribution in low-income countries', *Economic Development and Cultural Change*, vol. 27 (October 1978) no. 1.

2. The part of food produced for consumption by the household of the grower can be quite large. From 1973–4 to 1977–8 the gross marketable surplus of rice in Bangladesh ranged from 19 per cent of gross production to 23 per cent. In 1973–4 about 77 per cent of the gross marketable surplus came from 15 per cent of the farms. (See Raisuddin Ahmed, 'Agricultural rice policies . . .', 1981.)

3. This group is not as small as is often tacitly assumed. In Thailand one-fourth of paddy farmers are net purchasers of rice. It follows that the incidence of increases in the price of rice is quite different from what it would be if it could be assumed that all rice producers benefit from a price increase. (See Prasarn Trairatvorakul, 'The effects on income distribution and nutrition of alternative rice price policies in Thailand', International Food Policy Research Institute, Research Report, no. 46, November 1984). In Bangladesh about 53 per cent of farm households, covering 41 per cent of farm population and 19 per cent of farm land, were net buyers of rice. (See Raisuddin Ahmed, 'Agricultural price policies . . .', 1981.)

4. See pages 14–17.

5. Prasarn Trairatvorakul, 'Rice price policy and equity considerations in Thailand: distributional and nutritional effects' (Washington, DC: International Food Policy Research Institute) and 'The effects on income distribution and nutrition of alternative rice price policies in Thailand', International Food Policy Research Institute, Research Report, no. 46, November 1984.

13 Protecting the Poor in the Transition

1. A good discussion of some of the problems discussed in this section can be found in *Poverty and Hunger: Issues and Options for Food Security in Developing Countries*, a World Bank Policy Study, 1986.

2. See Ahmed, 'Agricultural price policies . . . .', note 12 to Chapter 5.

3. See Per Pinstrup-Anderson, 'Food subsidies: the concern to provide consumer welfare while assuring producer incentives', (IFPRI 1984), from which the reference in the next note is taken.

4. G. H. Beaton and H. Ghassemi, 'Supplementary feeding programmes for young children in developing countries', Report prepared for UNICEF and the ACC Committee on Nutrition, 1979.

14 Small Farmers and Employment

1. See Gustav Ranis and Frances Stewart, 'Rural Linkages in the Philippines and Taiwan', in *Macropolicies for Appropriate Technology*, forthcoming. I owe these points to Frances Stewart.

15 Rural–Urban Migration

1. This chapter has greatly benefited from comments by Oded Stark.

2. Malcolm D. Bale and Ernst Lutz, 'Price distortions in agriculture and their effects', World Bank Staff Working Paper, no. 359 (1979).

3. Tibor Scitovsky, 'Comment on Adelman', *World Development* (September 1984), and 'Economic development in Taiwan and South Korea', Stanford Food Research Institute, vol. XIX, no. 3 (1985) p. 231.

16 The Politics of Food Prices

1. Robert H. Bates, *Markets and States in Tropical Africa: The Political Basis of Agricultural Policies* (Berkeley: University of California Press, 1981); *Essays on the Political Economy of Rural Africa* (Cambridge: Cambridge University Press, 1983); and Michael Lipton, *Why Poor People Stay Poor; Urban Bias in World Development* (London: Temple Smith, 1977).
2. Mancur Olson, *The Logic of Collective Action* (Cambridge, Mass.: Harvard University Press 1965); *The Rise and Decline of Nations* (New Haven, Conn.: Yale University Press 1982); and 'The exploitation and subsidization of agriculture in developing and developed countries', mimeo. 1985. But, as Michael Lipton has pointed out, one would expect the government to respond to the political and economic weight of a pressure group, as well as to its coherence; and it is only the latter that is affected by the capacity of small groups to avoid the 'contributor's dilemma'.
3. R. H. Coase, 'The problem of social cost', *Journal of Law and Economics*, 3 (October 1960).
4. For a brilliant discussion of the relation between prisoners' dilemma and Coase's theorem in a different context, see Michael Lipton, 'Prisoners' Dilemma and Coase's Theorem: A Case for Democracy in Less Developed Countries?' in R.C.O. Matthews (ed.) *Economy and Democracy* (London: Macmillan, 1985).
5. See Michael Lipton, 'Prisoners' Dilemma and Coase's Theorem'.
6. Stephen Magee 'Endogenous Tariff Theory: A Survey', in *Neoclassical Political Economy: The Analysis of Rent-seeking and D.U.P. Activities*, edited by David C. Colander (Cambridge, Mass.: Ballinger, 1984).
7. As Manfred Bienefeld points out, at the end of his book *Markets and States in Tropical Africa*, Bates says: 'Alternatively, in response to the erosion of advantages engendered by shortfalls in production, the dominant interests may be persuaded to forsake the pursuit of unilateral short-run advantage, and instead to employ strategies that evoke cooperation by sharing joint gains' (IDS *Bulletin*, January 1986, p. 10). But once this possibility is envisaged, the simplicity and neatness of Bates' argument is removed and choices become more complex and uncertain.
8. Thomas Schelling, 'On the ecology of micromotives', *in The Public Interest* no. 25, Fall 1971; reproduced in *Micromotives and Macrobehavior*, W. W. Norton, 1978.
9. It might be thought that if high-income countries with a small proportion of farmers protect them at the expense of the large urban population, while poor countries, with a large agricultural population

and a small urban population protect the urban minority at the expense of the rural majority, there must be some income level and distribution of population in between which is just right and where no discrimination, protection and exploitation takes place. The situation is reminiscent of the man who complained to a mathematician friend that, when he was young, he always liked much older women and when he was old, he always liked young girls. The mathematician friend said: 'But, as you grew older and your tastes changed, there must have come a moment when they were the same age as you;' to which the man replied: 'Ah, but what is a moment!' Such a 'moment' must have occurred in South Korea when it switched from discriminating against agriculture to discriminating in its favour. In England, the 'moment' must have occurred in the late eighteenth or early nineteenth century.

10. 'The government of towns corporate was altogether in the hands of traders and artificers; and it was in the manifest interest of every particular class of them, to prevent the market from being overstocked, as they commonly express it, with their own particular species of industry, which is in reality to keep it always understocked . . . In their dealings with the country they were all great gainers . . . Whatever regulations . . . tend to increase those wages and profits beyond what they would otherwise be, tend to enable the town to purchase, with a smaller quantity of its labour, the produce of a greater quantity of the labour of the country. They give the traders and artificers of the town an advantage over the landlords, farmers, and labourers in the country, and break down the natural equality which would otherwise take place in the commerce which is carried on between them . . . The industry that is carried on in towns is . . . more advantageous than that which is carried on in the country . . . In every country of Europe we find, at least, a hundred people who have acquired great fortunes . . . for every one who has done so by . . . raising of rude produce by the improvement and cultivation of land.' Adam Smith, *The Wealth of Nations*, Book I, chapter X, part II. I owe the quotation to the paper by Mancur Olson.

11. Gary Anderson and Robert Tollison argue that the Factory Acts were not the result of humanitarian impulses but 'actually represented the mechanism by which skilled male operatives attempted to limit competition from alternative labour supplies'. ('A Rent-Seeking Explanation of the British Factory Acts', chapter 13 in *Neoclassical Political Economy; The Analysis of Rent-Seeking and DUP Activities*, edited by David C. Colander (Cambridge, Mass.; Ballinger, 1984.) If this view is accepted, it reinforces the argument in the text that interest differences within powerful groups can be used for the benefit of the oppressed.

12. See Manfred Bienefeld, 'Analysing the politics of African state policy: some thoughts on Robert Bates' work', *IDS Bulletin* vol. 17 (1986) no. 1.

13. Max Corden, 'The Normative Theory of International Trade', in Ronald W. Jones and Peter B. Kenen (eds), *Handbook of International Economics*, 1 (Amsterdam: North Holland, 1984).

17 The Budgetary Constraint

1. Lucio G. Reca, 'Argentina: country case study of agricultural prices, taxes and subsidies', World Bank Staff Working Paper, no. 386 (April 1980).

18 Balance of Payments Constraint

1. Malcolm D. Bale, 'Agricultural trade and food policy; the experience of five developing countries', World Bank Staff Working Paper, no. 724, (1985). See also Ramgopal Agarwala 'Price distortions and growth in developing countries', World Bank Staff Working Paper, no. 575 (1983).
2. Botswana, the country with a record rate of annual agricultural growth of 8.5 per cent between 1970 and 1981, had an appreciating currency and Ghana, the country with the highest rate of appreciation, did only enjoy zero, but not, like several others, negative agricultural growth (see p. 19 of Cleaver's paper).
3. Kevin M. Cleaver, 'The impact of price and exchange rate policies on agriculture in sub-Saharan Africa', World Bank Staff Working Paper, no. 728, (1985) p. 28. According to his calculation, a 1 per cent p.a. increase in the rate of currency depreciation is associated with a 0.15 per cent increase in agricultural growth: not a very large impact. Kevin Cleaver's paper suffers from some technical defects, but the general line remains interesting.

19 International Implications

1. Presidential Commission on World Hunger, *Overcoming World Hunger: The Challenge Ahead* (Washington, DC.: US Government Printing Office, March 1980) pp. 89–90; quoted in Robert Paarlberg 'Food Security Approach in the 1980s: Righting the Balance', *US Foreign Policy and the Third World Agenda 1982* (New York: Overseas Development Council, Praeger, 1982).

20 The Wider Issues of Hunger and Malnutrition

1. Shlomo Reutlinger and Marcelo Selowsky, *Malnutrition and Poverty: Magnitude and Policy Options* (Baltimore, MD: Johns Hopkins University Press, 1976). If the criticism that the authors overestimate malnutrition is valid, the proportion would be even less.
2. A. Edward Schuh, 'The world food situation', paper presented at the Seventh World Congress of the International Economic Associations, 1983, table 1.
3. To achieve perfect nutritional standards is virtually impossible, and a reasonable objective is to reduce the significant handicaps from nutritional deficiency. Many people in rich countries present medical problems from being overweight and obese, but no great social significance is attached to these ills. There may be as many as 7 million

Americans suffering from malnutrition. In poor countries, people have adapted to mild cases of calorie deficiency by attaining a lower weight and height, by being less active, and, in the case of women, by ovulating less regularly.

4.   Daniel R. Gross and Barbara A. Underwood, 'Technological change and caloric costs: sisal agriculture in northeastern Brazil', *American Anthropologist*, vol. 73, no. 3 (June 1971), pp. 725–40.

21 Recommendations

1.   A. K. Sen, 'Goods and people', paper presented to the Plenary Session of the Seventh World Congress of the International Economic Association in Madrid, 1983.

Index

124

Index